T0215358

Fundamentals of
Laser Optoelectronics
Second Edition

Series in Optics and Photonics

Series Editor: S L Chin *(Laval University, Canada)*

Series in Optics and Photonics – Vol. 9

Fundamentals of Laser Optoelectronics

Second Edition

See Leang Chin
Université Laval, Canada

Huailiang Xu
Jilin University, China

Shuai Yuan
University of Shanghai for Science and Technology, China

World Scientific

NEW JERSEY · LONDON · SINGAPORE · BEIJING · SHANGHAI · HONG KONG · TAIPEI · CHENNAI · TOKYO

Published by

World Scientific Publishing Co. Pte. Ltd.

5 Toh Tuck Link, Singapore 596224

USA office: 27 Warren Street, Suite 401-402, Hackensack, NJ 07601

UK office: 57 Shelton Street, Covent Garden, London WC2H 9HE

Library of Congress Cataloging-in-Publication Data
Names: Chin, See Leang, author. | Xu, Huailiang, author. |
    Yuan, Shuai (Professor of physics), author.
Title: Fundamentals of laser optoelectronics / See Leang Chin, Université Laval, Canada,
    Huailiang Xu, Jilin University, China,
    Shuai Yuan, University of Shanghai for Science and Technology, China.
Description: Second edition. | New Jersey : World Scientific, [2023] | Series: Series in optics and
    photonics, 2010-2313 ; Vol. 9 | Revised edition of: Femtosecond laser filamentation /
    See Leang Chin. c2010. | Includes bibliographical references and index.
Identifiers: LCCN 2022024378 | ISBN 9789811254758 (hardcover) |
    ISBN 9789811254987 (paperback) | ISBN 9789811254765 (ebook for institutions) |
    ISBN 9789811254772 (ebook for individuals)
Subjects: LCSH: Femtosecond lasers. | Laser pulses, Ultrashort.
Classification: LCC QC689.5.L37 C45 2023 | DDC 621.36/6--dc23/eng20220825
LC record available at https://lccn.loc.gov/2022024378

British Library Cataloguing-in-Publication Data
A catalogue record for this book is available from the British Library.

For any available supplementary material, please visit
https://www.worldscientific.com/worldscibooks/10.1142/12790#t=suppl

Desk Editors: Balasubramanian Shanmugam/Amanda Yun

Typeset by Stallion Press
Email: enquiries@stallionpress.com

Printed in Singapore

Foreword

Since the publication of the first edition of this book in 1989, developments based on laser systems and their applications have virtually exploded. A major factor behind such tremendous developments was the advent of modern high-power femtosecond lasers, including multi-petawatt (PW, 10^{15} W) laser systems. The latter is based upon the principle of chirped pulse amplification, the 2018 Nobel physics prize-winning technology invented by Gérard Mourou and Donna Strickland. Such technology revolutionized the whole industry behind high-power laser systems, which guaranteed lasers of both ultrahigh peak power and ultrashort pulse duration. High-power femtosecond lasers stimulated the development of micro science. The intense laser field is one of the best ways for atomic/electronic manipulations. At the same time, high-power femtosecond lasers also brought massive technologies to the world of industry, for instance, photolithography in integrated chips processing that has capacities of lithography under tens of nanometers. Chirped pulse amplification provides an alternative approach for the design of high-power femtosecond lasers. In fact, an understanding of the basic principle of chirped pulse amplification and instruments from modern laser science is almost a prerequisite for laser designers and scientists who work with laser systems.

This second edition of *Fundamentals of Laser Optoelectronics* brings us up to date. An attempt was made to include recent research results on modern technologies and instruments relevant to laser optoelectronics in each chapter. For instance, physical explanations for group velocity dispersion, group delay dispersion, and third-order dispersion are added in Chapter 1, an introduction of different types of laser systems is included in Chapter 4, and both optical isotropy and anisotropy in different types of harmonic generation are explained in Chapter 11. At the same time, this edition keeps most of the chapters that appeared in the first edition, except Chapter 12, which is now devoted to mode locking and carrier-envelope phase locking. We also added one more chapter (Chapter 13), which focuses on chirped pulse amplification. Theories based upon mode locking and chirped pulse amplifications have become increasingly more important. It is thus necessary that students learn all these in a course devoted to laser optoelectronics.

The book is based on a course given by the authors to third and fourth-year undergraduate students of physics, engineering physics, and electrical engineering. The purpose of the book is to introduce some of the fundamental principles underlying laser beam control in optoelectronics. This edition should continue to be a useful text for undergraduate courses. Graduated students and scientists working in the above-mentioned fields might also find their references during research works.

About the Authors

 **See Leang Chin** is Professor Emeritus of Physics at Laval University, Quebec City, Canada. He occupied the position of senior Canada Research Chair (CRC) in ultrafast intense laser sciences since the inception of the CRC program in 2000 by the federal government. He is a pioneer in intense laser multiphoton/tunnel ionization of atoms and molecules as well as ultrafast intense laser filamentation. In 2001, he received the Humboldt Research Award in Germany, the highest distinction of the Humboldt Research Foundation for a foreign scientist. In 2011, he received the highest award of the Canadian Association of Physicists, namely, the Medal of Lifetime Achievement in Physics. He is a fellow of the Optical Society of America (OSA now Optica). He has published more than 400 research papers.

 **Huailiang Xu** received his Ph.D degree in Physics from Lund University, Sweden, in 2004. He then worked as a postdoctoral researcher at the Department of Physics, Laval University, Canada. In January 2008, he became an Assistant Professor at the Department of Chemistry, University of Tokyo, Japan. Since September 2009, he has been a full professor at Jilin University, China. His research interests include

Laser spectroscopy and Strong Laser-matter interaction. He received National Distinguished Young Scholar from NSFC in 2016 and has published more than 180 papers in journals and five Springer book chapters.

Shuai Yuan is an Associate Professor at the Shanghai University of Science and Technology of China. He received his Ph.D degree in Physics from Laval University, Quebec City, Canada, in 2014. He is a Hope Fellow of Japan Society for the Promotion of Science (JSPS). He has published more than 30 papers in journals, 15 patents, and one book chapter. His publications have been cited over 400 times. His research interests include femtosecond laser science, filamentation, fiber optics, and remote sensing and monitoring.

Contents

Introduction

A laser is now a tool for many scientists and engineers. The application (or use) of lasers is so widespread that in almost every field of science, engineering and technology, including medicine and everyday life, we can find lasers in use. At the same time, new engineering and technological disciplines are created. Names such as photonics, optoelectronics, optronics, etc., are becoming normal vocabulary. Today, optical technology certainly plays a significant role in our lives. More and more people are going into the vast field of optics, once only narrowly defined as a lens maker's little world. Such a big change (or revolution) is in part due to the advancement in electronics but it is the invention of the laser 30 years ago that made this revolution possible. In fact, a laser is the heart of most optical, optoelectronic and photonic applications. There is thus a need among undergraduate students and newcomers to the field to have a book at hand that discusses the fundamental principles of controlling a laser beam. It is hoped that readers will have a good background so that if they were to go into any laser, optoelectronic or photonic laboratory, they can quickly learn to use the relevant equipment and control laser (and light) beams without difficulty.

Controlling laser (and light) beams means making any of the following types of changes (or manipulations) of a laser beam; several of them are interrelated.

(a) *Temporal change*: Temporal laser mode control, switching (ON/OFF), laser pulse slicing, etc.
(b) *Spatial energy (intensity) distribution change*: Spatial filtering, apodization, spatial laser mode control, etc.

(c) *Directional change*: Deviation by reflection, refraction and diffraction (grating, active or passive, etc.), and guiding (fibers and waveguides).
(d) *Total intensity (power or energy) change*: Amplitude modulation, switching, mode control, etc.
(e) *Frequency/phase change*: Laser mode control, phase (frequency) modulation, nonlinear optics, etc.
(f) *Polarization change*: Propagation through wave plates, retarders (active and passive), dichroic polarizers, etc.

Each manipulation has a certain purpose and application. Knowing the basic principles underlying the techniques of controlling the laser (light) beam will be a great help to all those who work with lasers and optical applications.

Three major subjects — optics of isotropic media, optics of anisotropic media, and laser — underlie most of the above-mentioned laser beam control techniques. This book explains the principles of the physical phenomena underlying the last two subjects at the undergraduate level, especially the part on optical anisotropy. This is because optical anisotropy is the heart of many electro-optic, magneto-optic, acousto-optic, nonlinear optical, Q-switching, and mode-locking devices. The principles of these latter devices are also treated. The optics of isotropic media is classical optics, and the reader is assumed to know about them to a certain extent. Whenever necessary, there will be a reminder about a particular subject. No attempt is made to deal with fiber optics. The book concludes with an introduction of chirped pulse amplification, which is the 2018 Nobel physics winning technology.

Problems for students are <u>not</u> explicitly given at the end of each chapter. Instead, they are given along the course of the text and are indicated by the underlined word "exercise".

Chapter 1

Maxwell's Equations, Wave Equation and Waves: A Review

1.1 A Pictorial View of EM Waves

The concept of the electromagnetic (EM) wave is central to this book. Though we talk about manipulation and control of laser (and light) beams, we need to use the wave concept to interpret propagation and interaction. Such a concept originates from Maxwell's equations. In general, they are given by:

$$\nabla \cdot \boldsymbol{E} = \rho/\varepsilon_0, \tag{1.1}$$

$$\nabla \cdot \boldsymbol{B} = 0, \tag{1.2}$$

$$\nabla \times \boldsymbol{E} = -\frac{\partial \boldsymbol{B}}{\partial t}, \tag{1.3}$$

$$\nabla \times \boldsymbol{B} = \mu_0 \varepsilon_0 \frac{\partial \boldsymbol{E}}{\partial t} + \mu_0 \boldsymbol{J}. \tag{1.4}$$

Equation (1.1) is the differential form of Coulomb's law where $\boldsymbol{E}$ is the electric field of a system of charges of density ρ, and ε_0 is the dielectric constant in free space, or the permittivity of vacuum.

$$\varepsilon_0 = 8.854 \times 10^{-12} F/m. \tag{1.5}$$

Equation (1.2) is the characteristic equation for a magnetic field $\boldsymbol{B}$, which does not have a point source, i.e., there is no magnetic monopole.

Equation (1.3) is the differential form of Faraday's law and Eq. (1.4) represents the modified Ampere law.

$$\mu_0 = 4\pi \times 10^{-7} H/m \tag{1.6}$$

is the permeability of vacuum and J is the current density.

In free space, $\rho = 0$, $J = 0$, and Eqs. (1.1) to (1.4) become:

$$\nabla \cdot E = 0, \tag{1.7}$$

$$\nabla \cdot B = 0, \tag{1.8}$$

$$\nabla \times E = -\frac{\partial B}{\partial t}, \tag{1.9}$$

$$\nabla \times B = \mu_0 \varepsilon_0 \frac{\partial E}{\partial t}. \tag{1.10}$$

Let us first of all interpret Eqs. (1.7) to (1.10) qualitatively. Mathematically speaking, when the divergence of a vector field is zero, i.e.,

$$\nabla \cdot V = 0,$$

where V is the vector field, it means that every field line will form a closed loop and they do not cross each other. From Eqs. (1.7) and (1.8), we can say that both the electric and magnetic field lines form closed loops in space, as shown schematically in Fig. 1.1(a) and 1.1(b). If there were only these two equations, then the E and B fields would be completely independent. Equations (1.9) and (1.10) link them together.

From Eq. (1.9), we understand that the time rate of change of the magnetic field generates a circulation of electric field around it (Faraday's law). This is shown pictorially in Fig. 1.1(c) where loops of $E(t)$ are generated at every point of the $B(t)$ loop. Only four points are indicated in Fig. 1.1(c). Note that the direction of E conforms with the negative sign in Eq. (1.9). The electric field, in turn, generates magnetic fields around it, according to Eq. (1.10), which says that the time rate of change of an electric field produces a circulation of magnetic field around it. This is shown in Fig. 1.1(d). We stress that because of the constraints of Eqs. (1.7) and (1.8), the electric and magnetic fields should form loops, as in Fig. 1.1(a) and (b). Also, the circulations in Eqs. (1.9) and (1.10) require them to form loops. Both these equations provide the coupling between them.

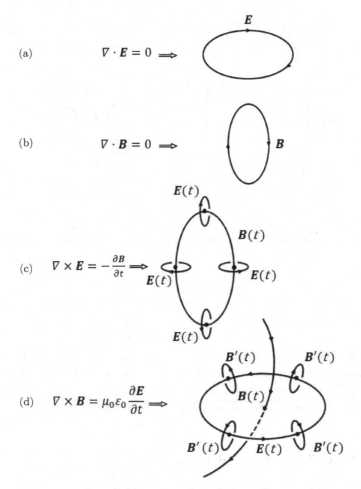

Fig. 1.1.   Pictorial interpretation of Maxwell's equations.

Now, the newly generated magnetic field (Fig. 1.1(d)) will, in turn, generate a new electric field around it, according to Eq. (1.9), and the new electric field will generate a newer magnetic field according to Eq. (1.10), and so on. Along one direction in space, the relationship between these successive generations of E and B field loops look like a chain, as shown in Fig. 1.2. Such "chain reactions" expand in all directions and this constitutes the propagation of the E and B fields. We now have a feeling of wave propagation. At every point P in space at time t, there is a resultant electric field and resultant magnetic

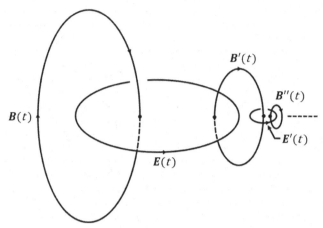

Fig. 1.2.   Pictorial visualization of electromagnetic field propagation by combining Maxwell's equations.

field. Each is a superposition (vector sum) of those expanding field loops that reach this space point P at time t. Such a *superposition principle* is an assumption that if we put a test charge (i.e., detector) at point P and time t, it will experience the resultant electric field mentioned above. Similarly, if we put a test current (i.e., detector) at P and t, we assume that it will experience the resultant $\boldsymbol{B}$. Such assumptions are proved to be valid experimentally.

1.2   Wave Equation and Plane Waves

A rigorous description of the fields $\boldsymbol{E}$ and $\boldsymbol{B}$ results from the solution of Maxwell's equations. We give a quick derivation of the wave equations. The following vector identity will be used:

$$\nabla \times \nabla \times \boldsymbol{V} = \nabla(\nabla \cdot \boldsymbol{V}) - (\nabla \cdot \nabla)\boldsymbol{V}, \qquad (1.11)$$

where $\boldsymbol{V}$ can be either $\boldsymbol{E}$ or $\boldsymbol{B}$. Because of Eqs. (1.7) and (1.8), we can simplify Eq. (1.11) to:

$$\nabla \times \nabla \times \boldsymbol{V} = -\nabla^2 \boldsymbol{V}. \qquad (1.12)$$

Now, taking the rotation of Eq. (1.9) gives:

$$\nabla \times \nabla \times \boldsymbol{E} = \nabla \times \left(-\frac{\partial \boldsymbol{B}}{\partial t} \right). \tag{1.13}$$

Using Eq. (1.12), Eq. (1.13) becomes:

$$-\nabla^2 \boldsymbol{E} = -\frac{\partial}{\partial t} \nabla \times \boldsymbol{B},$$

$$= -\frac{\partial}{\partial t} \left(\mu_0 \varepsilon_0 \frac{\partial \boldsymbol{E}}{\partial t} \right) \qquad \text{(from Eq. (1.10))}$$

$$\text{or} \quad \nabla^2 \boldsymbol{E} = \mu_0 \varepsilon_0 \frac{\partial^2 \boldsymbol{E}}{\partial t^2}. \tag{1.14}$$

Similarly, taking the rotation of Eq. (1.10) yields:

$$\nabla^2 \boldsymbol{B} = \mu_0 \varepsilon_0 \frac{\partial^2 \boldsymbol{B}}{\partial t^2}. \tag{1.15}$$

Equations (1.14) and (1.15) are the vector wave equations. As a reminder, a 1-D scalar wave equation is given (in the z-direction, say) mathematically by:

$$\frac{\partial^2 f}{\partial z^2} = \frac{1}{v^2} \frac{\partial^2 f}{\partial t^2}, \tag{1.16}$$

where f is the scalar wave function and v is the wave propagation velocity. Thus, comparing Eqs. (1.14) and (1.15) with (1.16), $\boldsymbol{E}$ and $\boldsymbol{B}$ each satisfies in the Cartesian coordinates, a 3-D vector wave equation with the same wave velocity

$$c \equiv \frac{1}{\sqrt{\mu_0 \varepsilon_0}}, \tag{1.17}$$

which is the velocity of light in free space.

Let us now use the 1-D scalar wave Eq. (1.16) to illustrate the nature of waves. Any function $f(z,t)$ of the type $f(t \pm z/v)$ or their linear combination is a solution to the wave Eq. (1.16). This can be shown by substitution into Eq. (1.16) (<u>exercise</u>). The functions $f_1(t-z/v)$ and $f_2(t+z/v)$ represent waves propagating in the positive and negative z-directions, respectively. This is just a mathematical consequence. We give a simple explanation. Consider the function

$$f(z,t) = f_1(t - z/v). \tag{1.18}$$

We keep in mind that z is position and t is time. Let the phase of $f(t - z/v)$ be some definite value, i.e.,

$$t - \frac{z}{v} = t', \tag{1.19}$$

where t' is any definite value. Equation (1.19) represents a family of parallel lines in the $z - t$ plane for different values of t' (Fig. 1.3(a)). Let us consider one of the lines given by Eq. (1.19). It means that for a fixed value of t', the function $f(z,t)$ of Eq. (1.18) will have the same value $f_1(t')$ whatever the changes in z and t are, so long as they are constrained by Eq. (1.19), i.e., so long as the combined changes of z and t follow Eq. (1.19), $f(z,t)$ will always have the same value $f_1(t')$.

Now, Eq. (1.19) can be re-written as:

$$z = v(t - t'). \tag{1.20}$$

When z and t changes according to Eq. (1.19) or (1.20), the rate of change of z with respect to t, from Eq. (1.20):

$$\frac{dz}{dt} = v. \tag{1.21}$$

Thus, the rate of change (velocity) is v and the direction of change is in the positive z-direction because t (time) always changes in the positive direction. Thus, the value $f_1(t')$ is constrained to move along the line given by Eq. (1.19) in the positive z-direction at the velocity v (see Fig. 1.3(b)). The reason why $f_1(t')$ should move at all is because t (time) changes continuously so z has to follow, according to Eq. (1.19). (Nothing will move if we could FREEZE t and z at some definite value.)

Similarly, another value of $f(z,t)$, say $f_1(t'')$, moves along the line

$$t - z/v = t'' \tag{1.22}$$

with the same velocity v in the same positive z-direction, and so on, i.e., every point of the function $f(z,t)$ given by Eq. (1.18) moves

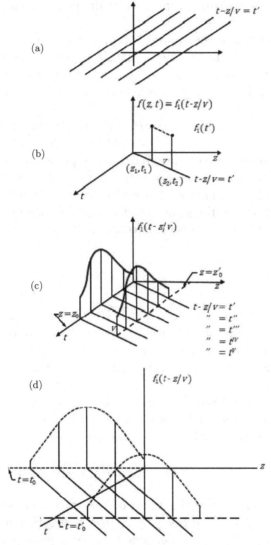

Fig. 1.3.   Pictorial visualization of 1-D wave propagation.

with the velocity v in the same positive z-direction without changing the form of the function. Figure 1.3(c) shows the propagation of the projection $f_1(t, z = z_0)$ moving from $z = z_0$ to $z = z_0'$ while Fig. 1.3(d) shows the propagation of the projection $f_1(t = t_0, z)$ moving from $t = t_0$ to $t = t_0'$. Similarly, the reader can analyze the motion of

$f(z, t) = f(t + z/v)$ in the negative z-direction. This is left as an exercise.

We now use only the spatial coordinate to describe the motion (Fig. 1.4(a)). Referring to Eq. (1.20), for a constant t', the wave function $f_1(t')$ moves along the z-axis. When the time changes from t_1 to t_2, $f(t')$, always a constant, moves from P_1 to P_2 with velocity v.

More generally, the distance OP_1 represents a plane perpendicular to the z-axis at P_1 (Fig. 1.4(b)) because $OP_1 = \boldsymbol{r} \cdot \hat{z}$ is the projection of the position vector $\boldsymbol{r}$ of every point of the plane on the z-axis. The equation of this plane is $\boldsymbol{r} \cdot \hat{z} = v(t_1 - t')(\equiv OP_1)$. When P_1 moves to P_2 at velocity v, plane $\boldsymbol{r} \cdot \hat{z} = v(t_1 - t')$ moves into plane $\boldsymbol{r} \cdot \hat{z} = v(t_2 - t') \equiv (OP_2)$ at the same velocity. $f_1(t')$ is constant everywhere on these two planes because t' is constant. Thus, we conclude that in general,

$$\boldsymbol{r} \cdot \hat{z} = v(t - t') \tag{1.23}$$

is a plane perpendicular to the z-axis. This plane moves in the positive z-direction at velocity v. At every point on this moving plane, the wave function $f_1(t')$ is a constant. We call the moving plane a <u>plane wavefront</u> and $f_1(t - z/v)$ a *plane wave function*.

The plane wavefront is a plane of constant phase (i.e., $t' =$ constant). For another value of the phase, say t'', the plane is shifted with respect to Eq. (1.23), i.e., the new plane is:

$$\boldsymbol{r} \cdot \hat{z} = v(t - t''),$$

and all over this plane, the value of the wave function is $f_1(t'')$, a constant. $f_1(t'')$ still propagates in the z-direction at velocity v (see Fig. 1.4(c)). Thus, we conclude that for a plane wave function $f_1(t - z/v)$, all the planes of constant phase

(i.e., $t - z/v = t' =$ constant or $t - \boldsymbol{r} \cdot \hat{z}/v = t' =$ constant)

propagate in the positive z-direction with velocity v. At every point on each moving plane, corresponding to one fixed phase t', the value of the wave function is the constant $f_1(t')$.

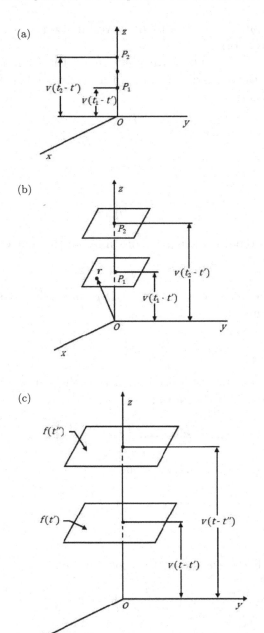

Fig. 1.4.   Pictorial visualization of 1-D plane wave propagation in space.

Similarly, $f_2(t + z/v)$ is a plane wave function propagating in the negative z-direction.

In optics, we often use wave functions of the form $f(\omega t \pm kz)$ or $f(\omega t \pm \boldsymbol{k} \cdot \boldsymbol{r})$. Both are of the same form as Eq. (1.18). For instance, we consider $f(\omega t - kz)$

$$\omega t - kz = \omega \left(t - \frac{z}{(\omega/k)} \right),$$

$$\equiv \omega(t - z/v), \tag{1.24}$$

if

$$v \equiv \omega/k. \tag{1.25}$$

For a monochromatic wave $\omega = $ constant, we thus have:

$$f(\omega t - kz) = f(\omega(t - z/v)),$$

which is of the same form as Eq. (1.18). Similarly, consider $f(\omega t - \boldsymbol{k} \cdot \boldsymbol{r})$, for example.

$$\omega t - \boldsymbol{k} \cdot \boldsymbol{r} = \omega \left(t - \frac{\boldsymbol{k} \cdot \boldsymbol{r}}{\omega} \right). \tag{1.26}$$

If we rotate the coordinate axes such that the z-axis coincides with the $\boldsymbol{k}$-axis, Eq. (1.26) becomes:

$$\omega t - \boldsymbol{k} \cdot \boldsymbol{r} = \omega \left(t - \frac{k\hat{z} \cdot \boldsymbol{r}}{\omega} \right). \tag{1.27}$$

Again, if $v \equiv \omega/k$, Eq. (1.27) becomes:

$$\omega t - \boldsymbol{k} \cdot \boldsymbol{r} = \omega \left(t - \frac{\boldsymbol{r} \cdot \hat{z}}{v} \right), \tag{1.28}$$

$$= \omega t', \tag{1.28'}$$

where we set Eq. (1.28) equal to a constant $\omega t'$

$$\omega \left(t - \frac{\boldsymbol{r} \cdot \hat{z}}{v} \right) = \omega t'$$

or

$$\boldsymbol{r} \cdot \hat{z} = v(t - t'), \tag{1.29}$$

which is identical to Eq. (1.23), i.e., $\omega t - \boldsymbol{k} \cdot \boldsymbol{r} = \omega t'$, a moving plane in the direction of $\boldsymbol{k}$ ($\hat{z}$ in Eq. (1.29)) at the velocity $v \equiv \omega/k$. On this plane, the wave function has a constant value $f_1(\omega t')$. A similar analysis applies to $f(\omega t + \boldsymbol{k} \cdot \boldsymbol{r})$.

In conclusion, whenever we see a function of the form (in *Cartesian coordinates*)

$$f(\omega t - \boldsymbol{k} \cdot \boldsymbol{r})$$

and

$$f(\omega t - kz), \text{etc.},$$

where ω = constant (monochromatic), they represent plane wave functions that are solutions to the wave equation. The plane wavefront of the wave function propagates in the direction of $\boldsymbol{k}$ (or $\hat{z}$) (i.e., $\boldsymbol{k}$ (or $\hat{z}$)) is the normal to the plane at the velocity ω/k. This velocity is called the *wave velocity*. On the plane wavefront, the wave function has a constant value. A similar argument applies to wavefronts propagating in the $(-k)$ direction.

In the case of the 3-D vector wave Eqs. (1.14) and (1.15), the plane wave solutions are of the form $\boldsymbol{E}(\omega t \pm \boldsymbol{k} \cdot \boldsymbol{r})$ and $\boldsymbol{B}(\omega t \pm \boldsymbol{k} \cdot \boldsymbol{r})$ in Cartesian coordinates with ω = constant. On the plane wavefront

$$\omega t \pm \boldsymbol{k} \cdot \boldsymbol{r} = \omega t', \quad (\text{see Eq. (1.28)})$$

the vector wave functions $\boldsymbol{E}$ and $\boldsymbol{B}$ are constant vectors. For simplicity in the rest of this chapter, we use scalar wave functions to describe different concepts of waves.

Customarily, in optics, the wave function is expressed in one of the following harmonic forms: $\cos(\omega t \pm \boldsymbol{k} \cdot \boldsymbol{r})$, $\sin(\omega t \pm \boldsymbol{k} \cdot \boldsymbol{r})$, $e^{\pm i(\omega t \pm \boldsymbol{k} \cdot \boldsymbol{r})}$, and the same functions, with $\boldsymbol{k} \cdot \boldsymbol{r}$ replaced by kz (or ky or kx).

1.3   Spherical Waves

If the wave function $f(rt)$ is a function of r and t only, where r is the radial position in space, it is convenient to express the 3-D wave equation in spherical coordinates (r, θ, ϕ). From vector calculus:

$$\nabla^2 f = \frac{1}{r^2} \frac{\partial}{\partial r} \left(r^2 \frac{\partial f}{\partial r} \right) + \frac{1}{r^2 \sin\theta} \frac{\partial}{\partial \theta} \left(\sin\theta \frac{\partial f}{\partial \theta} \right)$$

$$+ \frac{1}{r^2 \sin^2\theta} \frac{\partial^2 f}{\partial \phi^2}. \tag{1.30}$$

Since

$$f(\boldsymbol{r},t) = f(r,t), \text{ independent of } \theta \text{ and } \phi, \qquad (1.31)$$

Eq. (1.30) becomes:

$$\nabla^2 f = \frac{1}{r^2}\frac{\partial}{\partial r}\left(r^2\frac{\partial f}{\partial r}\right) \equiv \frac{1}{r}\frac{\partial^2}{\partial r^2}(rf). \qquad (1.32)$$

Substituting Eq. (1.32) into the 3-D wave equation $\nabla^2 f = \frac{1}{v^2}\frac{\partial^2 f}{\partial t^2}$, we have:

$$\frac{1}{r}\frac{\partial^2}{\partial r^2}(rf) = \frac{1}{v^2}\frac{\partial^2 f}{\partial t^2}$$

$$\text{or} \quad \frac{\partial^2}{\partial r^2}(rf) = \frac{1}{v^2}\frac{\partial^2(rf)}{\partial t^2}. \qquad (1.33)$$

Equation (1.33) is in the form of a 1-D wave equation whose wave function is (rf). Thus, the solution is:

$$rf(r,t) = g\left(t \pm \frac{r}{v}\right). \qquad (1.34)$$

Or in the form of $(\omega t - \boldsymbol{k}\cdot\boldsymbol{r})$, one has:

$$rf(r,t) = g(\omega t \pm kr), \quad \text{(independent of } \theta \text{ and } \phi\text{)}.$$

Hence,

$$f(r,t) = \frac{g(\omega t \pm kr)}{r}. \qquad (1.35)$$

In Eq. (1.35), $g(\omega t - kr)$ is a wave function whose wavefront is a spherical surface (note that we are now in the spherical coordinate; thus, $g(\omega t - kr)$ does not represent a plane wave) because it propagates with a velocity $v = \omega/k$ in all directions of r. The direction of propagation is that of increasing r, i.e., the spherical wavefront is diverging from the origin. Similarly, the wave function $g(\omega t + kr)$ has a spherical wavefront converging to the origin.

The total solution to the 3-D wave equation is the wave function $f(r,t)$ given by Eq. (1.35). It means that $f(r,t)$ are spherical wavefronts whose amplitude decreases linearly with r.

1.4 Wave Vector, Phase Velocity, Group Velocity

(A) We work now in Cartesian coordinates. The wave function is given by $f(\omega t - \boldsymbol{k} \cdot \boldsymbol{r})$.

<u>Def.</u> wave vector $\equiv \boldsymbol{k}$.

$\boldsymbol{k}$ is always perpendicular to the plane wavefront (Section 1.2, Eq. (1.28)).

$$\omega t - \boldsymbol{k} \cdot \boldsymbol{r} = \text{constant} \tag{1.36}$$

(The same orthogonal relationship applies to a spherical wavefront or any wavefront.)

Since the propagation velocity of the wavefront is the wave velocity,

$$v = \frac{\omega}{k}, \quad \left(k \equiv |\vec{k}|\right),$$

we have:

$$k = \frac{\omega}{v}. \tag{1.37}$$

In optics, when the wave $f(\omega t - \boldsymbol{k} \cdot \boldsymbol{r})$ propagates in a dielectric medium,

$$v = \frac{c}{n}, \tag{1.38}$$

where n is the refractive index. Substituting Eq. (1.38) into (1.37) yields

$$k = \frac{\omega}{c} n, \quad (\text{in a dielectric}). \tag{1.39}$$

If the wave is in vacuum, $n = 1$,

$$k_0 = \frac{\omega}{c}, \quad (\text{in vacuum}). \tag{1.40}$$

(B) When two monochromatic waves of slightly different frequencies come together, we can combine them using the *superposition principle*. This principle requires that the wave equation be linear. (Linear means that the wave function f and its derivatives all appear to the first order; there is no term in the wave equation containing $f^2, \left(\frac{\partial f}{\partial z}\right)^2, \ldots$ etc.)

Then any linear combination of all different $f's$ is also a solution.

Assume that the two waves (E_1, E_2) with frequencies ω_1 and ω_2 have the same amplitude. Assume that they also propagate in the same direction z.

$$E_1 = E_0 e^{-\mathrm{i}(\omega_1 t - k_1 z)}, \tag{1.41}$$

$$E_2 = E_0 e^{-\mathrm{i}(\omega_2 t - k_2 z)}. \tag{1.42}$$

Their superposition yields the total field, which is the summation of the two.

$$E = E_1 + E_2. \tag{1.43}$$

We define

$$\bar{\omega} \equiv \frac{1}{2}(\omega_1 + \omega_2), \tag{1.44}$$

$$\Delta\omega \equiv \omega_1 - \omega_2, \tag{1.45}$$

$$\bar{k} \equiv \frac{1}{2}(k_1 + k_2), \tag{1.46}$$

$$\Delta k \equiv k_1 - k_2. \tag{1.47}$$

Substituting Eqs. (1.44 to 1.47) and (1.41 and 1.42) into Eq. (1.43) yields (the reader can also do this exercise in the calculation):

$$E = 2E_0 e^{-\mathrm{i}(\bar{\omega}t - \bar{k}z)} \cos\left(\frac{\Delta\omega}{2}t - \frac{\Delta k z}{2}\right). \tag{1.48}$$

There are now two propagating wave functions in Eq. (1.48): $e^{-\mathrm{i}(\bar{\omega}t - \bar{k}z)}$ and $\cos\left(\frac{\Delta\omega}{2}t - \frac{\Delta k z}{2}\right)$. The former propagates with a mean frequency $\bar{\omega}$ and mean wave vector $\bar{k}$ so that its velocity is:

$$\bar{v}_{\mathrm{p}} = \bar{\omega}/\bar{k}, \tag{1.49}$$

while the second with a frequency $\frac{\Delta\omega}{2}$ and wave vector $\frac{\Delta k}{2}$ at a velocity

$$v_{\mathrm{g}} = \frac{\Delta\omega}{\Delta k}. \tag{1.50}$$

Special condition:
$$\left. \begin{array}{l} \Delta\omega/\bar{\omega} \ll 1 \\ \Delta k/\bar{k} \ll 1 \end{array} \right\}. \tag{1.51}$$

Under the condition (1.51), the frequency of the cosine term in Eq. (1.48) is much lower than the mean frequency $\overline{\omega}$ of the exponential term. We thus interpret Eq. (1.48) as having a carrier plane wave of mean frequency $\overline{\omega}$ (from $e^{-i(\overline{\omega}t-\overline{k}z)}$) whose amplitude is also a wave $\left(\cos\left(\frac{\Delta\omega}{2}t - \frac{\Delta kz}{2}\right)\right)$ propagating with a much lower *modulating* frequency $\Delta\omega/2$. The propagation velocity of the carrier wave is the *phase velocity* v_{p} (Eq. (1.49)) and that of the modulating amplitude is the *group velocity* v_{g} (Eq. (1.50)).

(C) When a group of monochromatic plane waves traveling in the same z-direction are superimposed together, the total field is the summation of them all. If their frequencies differ continuously across width $\Delta\omega$, the summation becomes an integral:

$$E(z,t) = \int_{\Delta\omega} E_0(\omega)e^{-i(\omega t-kz)}\,d\omega,$$

$$= \int_{\Delta\omega} E_0(\omega)e^{-i(\overline{\omega}t-\overline{k}z)}e^{-i(\omega-\overline{\omega})t}$$

$$\cdot e^{i(k-\overline{k})z}\,d\omega, \tag{1.52}$$

where $\overline{\omega}$ and $\overline{k}$ are the mean frequency and wave vector, respectively, and $\Delta\omega$ denotes the frequency interval around $\overline{\omega}$. Equation (1.52) can be regrouped:

$$E(z,t) = A(z,t)e^{-i(\omega t-kz)}, \tag{1.53}$$

where

$$A(z,t) \equiv \int_{\Delta\omega} E_0(\omega)e^{-i[(\omega-\overline{\omega})t-(k-\overline{k})z]}\,d\omega$$

$$\equiv \int_{\Delta\omega} E_0(\omega)e^{-i\left[(\omega-\overline{\omega})\left(t-\dfrac{k-\overline{k}}{\omega-\overline{\omega}}z\right)\right]}\,d\omega. \tag{1.54}$$

Again, as in the case of Eq. (1.48), we interpret Eq. (1.53) as having a carrier plane wave ($e^{-i(\overline{\omega}t-\overline{k}z)}$) whose plane of constant phase propagates in the z-direction at the wave velocity of $\overline{\omega}/\overline{k}$ and variable amplitude wave $A(z,t)$ given by Eq. (1.54).

It is easier to interpret the latter by rewriting it as:

$$A(z,t) = \int_{\Delta\omega} E_0(\omega) e^{-i[\omega_g(t - \frac{z}{v_g})]} d\omega, \qquad (1.55)$$

where

$$\omega_g = \omega - \overline{\omega}, \qquad (1.56)$$

$$v_g = \frac{\omega - \overline{\omega}}{k - \overline{k}}. \qquad (1.57)$$

Equation (1.55) reveals itself as a superposition of a group of wave functions of frequencies ω_g. On the surface:

$$t - \frac{z}{v_g} = 0 \qquad (1.58)$$

(cf. discussion around Eq. (1.23)) $A(z,t)$ is maximum everywhere. This plane surface of maximum $A(z,t)$ advances at the group *velocity* v_g, given by Eq. (1.57).

If $\qquad\qquad\qquad \Delta\omega/\overline{\omega} \ll 1, \Delta k/\overline{k} \ll 1,$

$$v_g = \frac{\omega - \overline{\omega}}{k - \overline{k}} \approx \frac{d\omega}{dk}. \qquad (1.59)$$

This is true for most spectral lines and laser pulses in which the width $\Delta\omega$ is narrow.

Thus, v_g is a function of k. Physically, this is already explained as being due to the dispersion, which means a different propagating velocity (or index) for a different frequency in a medium. From Eq. (1.39):

$$k = \frac{\omega}{c} n(\omega) \qquad (1.60)$$

because dispersion means that the refractive index n is a function of ω. Because of Eq. (1.60), v_g is also a function of frequency. Substituting Eq. (1.60) into (1.59), one can obtain:

$$v_g = \frac{d\omega}{dk} = \frac{1}{\left[\dfrac{dk}{d\omega}\right]} = \frac{1}{\left[\dfrac{n + \omega\frac{dn}{d\omega}}{c}\right]},$$

$$= \frac{c}{\left[\left(n + \omega\dfrac{dn}{d\omega}\right)\right]} = \frac{c}{n\left(1 + \dfrac{\omega}{n}\dfrac{dn}{d\omega}\right)} = v_p \Big/ \left(1 + \frac{\omega}{n}\frac{dn}{d\omega}\right).$$

$$(1.61)$$

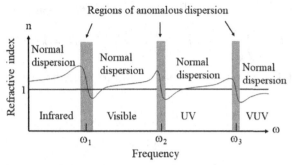

Fig. 1.5. Refractive index versus light frequency. The regions for the normal and anomalous dispersions are shown.

Therefore, when $dn/d\omega = 0$, the group velocity v_g equals the phase velocity v_p such as in a vacuum. Otherwise, when n increases with ω, i.e., $dn/d\omega > 0$, we have a normal (positive) dispersion or $v_g < v_p$. In some cases, for example, in an absorptive medium, as shown in Fig. 1.5, when $\omega \sim \omega_i$, where $\omega_i(i = 1, 2, 3)$ is the absorption frequency, the light at ω falls in the regions of anomalous dispersion, i.e., we have a negative dispersion of $dn/d\omega < 0$, $v_g > v_p$.

In addition, one often considers the refractive index n in terms of the wavelength λ. In this case, according to $\lambda = \frac{2\pi c}{\omega}$ and $\frac{d\lambda}{d\omega} = -\frac{\lambda^2}{2\pi c}$, Eq. (1.61) becomes:

$$v_g = \frac{c}{n\left(1 - \dfrac{\lambda dn}{n d\lambda}\right)}. \tag{1.62}$$

1.5 Group Velocity Dispersion and Group Delay Dispersion

In the case of ultrashort pulses in which $\Delta\omega$ is no longer much smaller than $\overline{\omega}$, we need to reconsider v_g. In fact, v_g is not a constant because of the dispersion of the different frequency components. Thus, all the waves propagate at different velocities, giving rise to what one calls "group velocity dispersion (GVD)".

In view of GVD, we need to consider a more general expression for the group velocity v_g. "Inspired" by Eqs. (1.57) and (1.59), we can, in general, make the following Taylor's expansion:

$$\omega = \overline{\omega} + \left(\frac{d\omega}{dk}\right)_{k=\overline{k}}(k - \overline{k}) + \frac{1}{2}\left(\frac{d^2\omega}{dk^2}\right)_{k=\overline{k}}(k - \overline{k})^2 + \cdots \quad (1.63)$$

We see that Eq. (1.63) gives the normal group velocity v_g (Eq. 1.59) if

$$|k - \overline{k}| \ll \overline{k} \quad (1.64)$$

or $$\Delta k \ll \overline{k},$$

which is the condition for the validity of Eq. (1.59). Under this condition, Eq. (1.63) becomes (by keeping the first two terms on the right hand side):

$$\omega = \overline{\omega} + \left(\frac{d\omega}{dk}\right)_{k=\overline{k}}(k - \overline{k}),$$

$$v_g \equiv \frac{\omega - \overline{\omega}}{k - \overline{k}} \approx \left(\frac{d\omega}{dk}\right)_{k=\overline{k}}. \quad (1.65)$$

If Eq. (1.63) is not valid, i.e., if $|k - \overline{k}|$ becomes larger, we need to consider the higher order terms in Eq. (1.63). If $|k - \overline{k}|$ and $|\omega - \overline{\omega}|$ are still not very large, adding the 3rd term is sufficient.

$$\omega = \overline{\omega} + \left(\frac{d\omega}{dk}\right)_{k=\overline{k}}(k - \overline{k}) + \frac{1}{2}\left(\frac{d^2\omega}{dk^2}\right)_{k=\overline{k}}(k - \overline{k})^2.$$

We have:

$$v_g \equiv \frac{\omega - \overline{\omega}}{k - \overline{k}} = \left(\frac{d\omega}{dk}\right)_{k=\overline{k}} + \frac{1}{2}\left(\frac{d^2\omega}{dk^2}\right)_{k=\overline{k}}(k - \overline{k}). \quad (1.66)$$

Let

$$v_{g0} \equiv \left(\frac{d\omega}{dk}\right)_{k=\overline{k}}, \quad (1.67)$$

$$\equiv \text{constant}.$$

Equation (1.66) becomes:

$$v_g \equiv v_{g0} + \frac{1}{2}\left(\frac{d^2\omega}{dk^2}\right)_{k=\overline{k}}(k - \overline{k}). \quad (1.68)$$

Thus, v_g is a function of k. Because of Eq. (1.60), v_g is also a function of frequency.

Definition: Group velocity spread (Δv_g)

From Eq. (1.68):

$$\Delta v_g = v_g - v_{g0} = \frac{1}{2}\left(\frac{d^2\omega}{dk^2}\right)_{k=\bar{k}}(k-\bar{k}). \qquad (1.69)$$

In the case of the propagation of an ultrashort laser pulse (ps to fs) in a medium of index $n(\omega)$, Eq. (1.69) tells us that there is a spread in the pulse's spatial extent of the order of $(\Delta v_g)t$, where t is the time of propagation.

If one considers the dispersion of the different frequency components during the propagation, the term $\left(\frac{d^2\omega}{dk^2}\right)_{k=\bar{k}}$ shows the distorted extent of the pulse envelope during its propagation. Thus, GVD can be expressed as:

$$k'' = 1/\left(\frac{d^2\omega}{dk^2}\right)_{k=\bar{k}} = \left(\frac{d^2k}{d\omega^2}\right)_{\omega=\bar{\omega}} = \frac{d}{d\omega}\left(\frac{1}{v_g(\omega)}\right)_{\omega=\bar{\omega}}, \qquad (1.70)$$

which represents the variation in group velocity with frequency; that is, GVD means that the group velocity will be different for different frequencies in the pulse.

According to $\frac{d\lambda}{d\omega} = -\frac{\lambda^2}{2\pi c}$ and Eq. (1.62), GVD in terms of wavelength λ can be expressed as:

$$k'' = \frac{d}{d\omega}\left(\frac{1}{v_g(\omega)}\right)_{\omega=\bar{\omega}} = \left(\frac{\lambda^3}{2\pi c^2}\frac{d^2n}{d\lambda^2}\right)_{\lambda=\bar{\lambda}}, \qquad (1.71)$$

which has the unit of s^2/m.

When a pulse propagates over a length L, the pulse envelope is delayed by the amount of $T_g(\bar{\omega})$, which is defined as the group delay.

$$T_g(\bar{\omega}) = \frac{L}{v_{g0}} = L/\left(\frac{d\omega}{dk}\right)_{k=\bar{k}}. \qquad (1.72)$$

Then, the variation in a group delay with frequency, called group delay dispersion (GDD), can be expressed as:

$$\text{GDD} = \text{GVD} \cdot L = k''L = L\frac{d}{d\omega}\left(\frac{1}{v_g(\omega)}\right)_{\omega=\bar{\omega}} = L\left(\frac{\lambda^3}{2\pi c^2}\frac{d^2n}{d\lambda^2}\right)_{\lambda=\bar{\lambda}}, \qquad (1.73)$$

which has the unit of fs^2 or fs/Hz.

Thus, the GDD results in a separation of frequency components of a pulse. For positive GDD, which is the most common situation such as in an ordinary transparent optical glass and crystal in the visible range, the components with shorter wavelengths are delayed with respect to those with longer wavelengths, giving rise to the longer wavelengths in the leading edge and the shorter wavelengths in the trailing edge of the pulse, i.e., a positive "chirp" of the pulse. Similarly, for negative GDD, the components with longer wavelengths are delayed with respect to those with shorter wavelengths, resulting in the shorter wavelengths in the leading edge and the longer wavelengths in the trailing edge of the pulse, i.e., a negative chirp of the pulse.

1.6   Third-Order Dispersion

If $|k - \overline{k}|$ and $|\omega - \overline{\omega}|$ are very large, such as in few-cycle pulses with very broad bandwidth, the first three terms in Eq. (1.63) will not be sufficient to express the frequency dispersion. In this case, higher orders of dispersion that generate more complicated distortions of a pulse shall also be considered. For example, for third-order dispersion (TOD), we have to make the following Taylor's expansion:

$$\omega = \overline{\omega} + \left(\frac{d\omega}{dk}\right)_{k=\overline{k}}(k - \overline{k}) + \frac{1}{2}\left(\frac{d^2\omega}{dk^2}\right)_{k=\overline{k}}(k - \overline{k})^2$$
$$+\frac{1}{6}\left(\frac{d^3\omega}{dk^3}\right)_{k=\overline{k}}(k - \overline{k})^3 \dots \qquad (1.74)$$

Let

$$k''' = 1/\left(\frac{d^3\omega}{dk^3}\right)_{k=\overline{k}}. \qquad (1.75)$$

We then have:

$$v_{\mathrm{g}} \equiv \frac{\omega - \overline{\omega}}{k - \overline{k}} = v_{\mathrm{g}0} + \frac{1}{2k''}(k - \overline{k}) + \frac{1}{6k'''}(k - \overline{k})^2 \qquad (1.76)$$

Similar to GDD, TOD can be expressed as:

$$\mathrm{TOD} = k'''L = L\left(\frac{dk^3}{d^3\omega}\right)_{\omega=\overline{\omega}} = L\frac{d^2}{d\omega^2}\left(\frac{1}{v_{\mathrm{g}}(\omega)}\right)_{\omega=\overline{\omega}}. \qquad (1.77)$$

It should be emphasized that in practice, the third-order dispersion term would only be considered when the pulse duration is short, typically less than 30 fs. While the second-order dispersion term will be considered widely for pulses with a duration in the picosecond to femtosecond range, we shall not go into the details here on how a short laser pulse behaves when propagating in a medium with normal or anomalous dispersions. Examples for setups to treat various pulse dispersions will be given in Chapter 13 when we introduce chirped pulse amplification.

1.7   Closing Remarks

We tried to give an elementary look at the wave nature of the electromagnetic fields E and B starting from Maxwell's equations. The pictorial descriptions of the propagation of E and B fields and part of the wave propagation represent the authors' "unorthodox" way of interpretation. Hopefully, they will give the reader a better feeling. More rigorous mathematical descriptions can be found in advanced texts, e.g., (Born and Wolf, 1980).

Reference

M. Born and E. Wolf (1980). *Principles of Optics*, 6th edition (Pergamon Press), Oxford.

Chapter 2

Snell's Law, Fresnel Equations, Brewster Angle and Critical Angle

We now consider how the Brewster angle window works and what makes the end prism totally reflecting. We only need to consider the idealized case of sending a plane monochromatic electromagnetic wave, say a well-collimated laser beam, across an interface separating two different isotropic materials, as shown in Fig. 2.1. The medium in which the incident beam propagates is assumed transparent to the particular frequency of the EM wave. The second medium could either be a transparent one or a reflecting one. But let us now consider only the case in which the second medium is also transparent.

2.1 Reflection and Refraction at Boundaries

Considering only the boundary conditions under which an electromagnetic wave crosses an interface separating two isotropic media, i.e.,

(i) The tangential components of the electric field E are continuous,
(ii) The tangential components of B/μ (or H) are continuous, one can derive the following relationships (cf. any optics text).

(1) The relations that show the directions of the reflected and transmitted beams, i.e.,

(a) The incident, reflected and transmitted beams are in the same plane, called the plane of incidence,

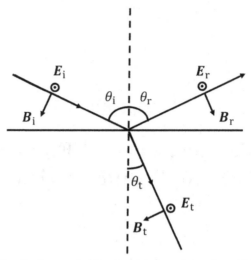

Fig. 2.1.   Fresnel reflection and refraction across a boundary. The electric vector of the incident wave is perpendicular to the plane of incidence.

(b) $\theta_i = \theta_r,$ and   (2.1)

(c) $n_i \sin(\theta_i) = n_t \sin(\theta_t).$ (2.2)

The second relation (Eq. (2.2)) is called <u>Snell's</u> law of refraction.
(2) The relations that give the amplitudes of the reflected and trans-
     mitted electric fields relative to the incident electric field, i.e.,
     <u>Fresnel's</u> equations. They are:

$$r_\perp \equiv \left(\frac{E_{Or}}{E_{Oi}}\right)_\perp = \frac{n_i \cos\theta_i - n_t \cos\theta_t}{n_i \cos\theta_i + n_t \cos\theta_t} = -\frac{\sin(\theta_i - \theta_t)}{\sin(\theta_i + \theta_t)}, \qquad (2.3)$$

$$r_\parallel \equiv \left(\frac{E_{Or}}{E_{Oi}}\right)_\parallel = \frac{n_t \cos\theta_i - n_i \cos\theta_t}{n_i \cos\theta_t + n_t \cos\theta_i} = \frac{\tan(\theta_i - \theta_t)}{\tan(\theta_i + \theta_t)}, \qquad (2.4)$$

$$t_\perp \equiv \left(\frac{E_{Ot}}{E_{Oi}}\right)_\perp = \frac{2n_i \cos\theta_i}{n_i \cos\theta_i + n_t \cos\theta_t} = \frac{2\sin\theta_t \cos\theta_i}{\sin(\theta_i + \theta_t)}, \qquad (2.5)$$

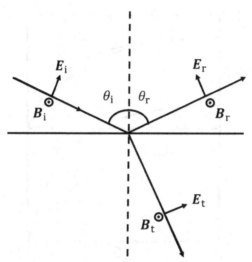

Fig. 2.2.   Fresnel reflection and refraction across a boundary. The electric vector of the incident wave is in the plane of incidence.

$$t_\| \equiv \left(\frac{E_{Ot}}{E_{Oi}}\right)_\| = \frac{2n_i \cos\theta_i}{n_i \cos\theta_t + n_t \cos\theta_i} = \frac{2\sin\theta_t \cos\theta_i}{\sin(\theta_i + \theta_t)\cos(\theta_i - \theta_t)},$$

$$(2.6)$$

where E_{Oj} ($j = $ i,r,t) is the amplitude of the electric field, and $\perp(\|)$ means the incident $\boldsymbol{E}$ field is perpendicular (parallel) to the plane of incidence as shown in Fig. 2.1 (see also Fig. 2.2). We assume that the reflected and transmitted electric fields are in the directions shown in Figs. 2.1 and 2.2. The Fresnel equations will correct the situation if the assumed direction is not right. For instance, Eq. (2.3) shows that $r_\perp$ is negative. It means that the amplitude E_{Or} changes sign with respect to the incident field. The monochromatic fields of the incident (i), reflected (r) and transmitted (t) fields are:

$$\boldsymbol{E}_i = \boldsymbol{E}_{Oi} \cos(\boldsymbol{k}_i \cdot \boldsymbol{r} - \omega t), \qquad (2.7)$$

$$\boldsymbol{E}_r = \boldsymbol{E}_{Or} \cos(\boldsymbol{k}_r \cdot \boldsymbol{r} - \omega t + \epsilon_r), \qquad (2.8)$$

$$\boldsymbol{E}_t = \boldsymbol{E}_{Ot} \cos(\boldsymbol{k}_t \cdot \boldsymbol{r} - \omega t + \epsilon_t), \qquad (2.9)$$

where ϵ_r and ϵ_t are constant phases with respect to the incident field. The proof of the Fresnel equations is straightforward and

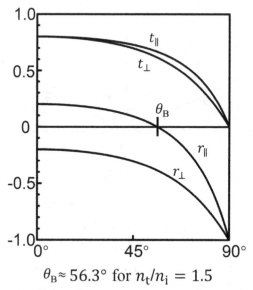

$$\theta_B \approx 56.3° \text{ for } n_t/n_i = 1.5$$

Fig. 2.3. Schematic relations between $r_\perp$, $r_\parallel$, $t_\perp$ and $t_\parallel$, and the angle of incidence (horizontal scale) for a material boundary in which $n_t/n_i = 1.5$.

can be found in any optics text, and we shall not derive them here. We only want to point out some useful features of these equations. Figure 2.3 shows four curves representing the variation of $r_\perp$, $r_\parallel$, $t_\perp$ and $t_\parallel$, as one changes θ_i under the condition that $\underline{n_i < n_t}$, where the **n**'s are the indices of refraction. Note that the t's are positive, while $r_\perp$ is negative and $r_\parallel$ starts from a positive value, passes through zero, and then becomes negative.

What is interesting here is the following. (Note, first of all, the definition of r's and t's in Eqs. (2.3) to (2.6).)

(a) $r_\perp <$ (Eq. (2.3)) for all θ_i and θ_t, i.e., the reflected electric field amplitude is 180° out of phase with respect to the incident electric field amplitude:

$$(\because -1 = e^{\pm i\pi}).$$

(b) $r_\parallel = 0$ at $\theta_i = \theta_B$, where θ_B is defined as the polarizing angle or Brewster angle.

From Eq. (2.4), $r_\| = 0$ means
either

$$\tan(\theta_i - \theta_t) = 0, \tag{2.10}$$

or

$$\tan(\theta_i + \theta_t) = \infty. \tag{2.11}$$

Equation (2.10) is impossible because $\theta_i \neq \theta_t \neq$ in general. Thus Eq. (2.11) has to be valid, i.e.,

$$\theta_i + \theta_t = \pi/2,$$

or

$$\theta_B + \theta_t = \pi/2, \quad (\therefore \ \theta_i = \theta_B). \tag{2.12}$$

Using Snell's law (Eq. (2.2)):

$$n_i \sin\theta_B = n_t \sin\theta_t,$$

$$= n_t \sin\left(\frac{\pi}{2} - \theta_B\right),$$

$$= n_t \cos\theta_B,$$

$$\therefore \ \tan\theta_B = \frac{n_t}{n_i}. \tag{2.13}$$

This is an important result that is used very often by optics or laser users. (Note: The relation is valid also in the case of refraction from a denser medium, i.e., $n_i > n_t$. The angle θ_B' in this case is such that $\theta_B + \theta'_B = \frac{\pi}{2}$.)

However, one cannot experimentally measure the field E. Rather it is the <u>power</u> (Joules/sec) of the radiation that one measures. The power is related to the <u>intensity</u> or <u>irradiance</u> (Joules/m^2 sec) by a simple factor, the cross-sectional area of the beam of radiation. The intensity is the averaged value of the Poynting vector, defined in electromagnetic theory as:

$$\boldsymbol{S} = \boldsymbol{E} \times \boldsymbol{B}/\mu. \tag{2.14}$$

Thus, referring to Fig. 2.4, where a beam of light of finite cross-section is reflected and transmitted at an interface, one can prove the following using the Fresnel equations.

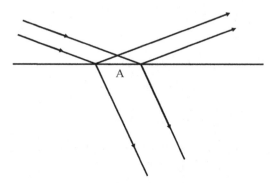

Fig. 2.4.   Reflectance and transmittance across a boundary.

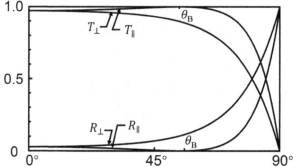

Fig. 2.5.   Variation of $R_\perp$, $R_\parallel$, $T_\perp$ and $T_\parallel$ as a function of the angle of incidence θ_i (horizontal scale).

(a) The reflectance R

$$\equiv \frac{\text{reflected power}}{\text{incident power}} = \frac{E_{\text{Or}}^2}{E_{\text{Oi}}^2} = r^2. \qquad (2.15)$$

(b) The transmittance T

$$= \frac{\text{transmitted power}}{\text{incident power}},$$

$$= \left(\frac{n_t \cos \theta_t}{n_i \cos \theta_i}\right) t^2. \qquad (2.16)$$

(c) $$R + T = 1, \qquad (2.17)$$

which is the law of conservation of energy, assuming no other loss exists at the boundary.

Figure 2.5 shows the variation of $R_\perp$, $R_\parallel$, $T_\perp$ and $T_\parallel$ as a function of incident angle θ_i for $n_t > n_i$. Note again $R_\parallel = 0$ at $\theta_i = \theta_B$, at the Brewster angle. No electromagnetic power can be reflected at the Brewster angle.

2.2 Taking Advantage of the Brewster Angle and the Features of the Reflectance and Transmittance

(a) Since at $\theta_i = \theta_B$ with the reflectance $R = 0$, one often sets the transmitting window of a linearly polarized laser beam at the Brewster angle so that the polarization of the laser is in the plane of incidence as shown in Fig. 2.6. This eliminates the loss by reflection so that almost all the energy in the laser beam is transmitted. (There are still scattering, absorption and diffraction losses.)

(b) By stacking a good number of plane-parallel plates at the Brewster angle, an unpolarized laser beam can be made polarized by passing through this stack, as shown in Fig. 2.7. All the radiation

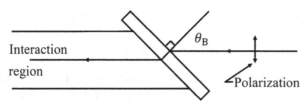

Fig. 2.6. Transmission through a window set at the Brewster angle θ_B. The polarization (linear) of the incident wave is in the plane of incidence.

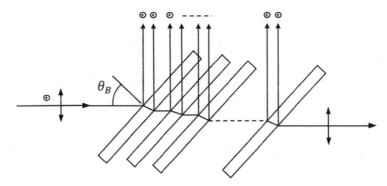

Fig. 2.7. A stack of windows set at the Brewster angle acts as a polarizer.

polarized (or whose electric field is) perpendicular to the plane of incidence will be successively partially reflected, while the other component parallel to the plane of incidence will be transmitted without any reflection. Thus, after many such passages through the stack, the transmitted beam is polarized linearly in the plane of incidence to a good degree of accuracy. We define the degree D of linear polarization as:

$$D = \frac{P_\parallel}{P_\perp}, \qquad (2.18)$$

where $P_\parallel \, (P_\perp)$ is the power of the radiation in the same laser beam whose polarization is parallel (perpendicular) to the plane of incidence.

Let us try to make a simple calculation. Assuming that in the incident laser beam, $P_\parallel^{(in)} = P_\perp^{(in)} \equiv P_0$. After transmitting through one plate set at the Brewster angle, $P_\parallel$ is unchanged while there have been two reflections at both the interfaces of the plate. Note that the two $R_\perp$'s at the two interfaces are both equal to $\sin^2(\theta_B - \theta_t)$ at the Brewster angle, as can be seen from Eq. (2.3) and the condition $\theta_i + \theta_t = \pi/2$ at $\theta_i = \theta_B$. Hence, the power of the $\perp$ polarization in the transmitted beam $P_\perp^{(1 \text{ plate})}$ is:

$$P_\perp^{(1\,\text{plate})} = P_0(1 - R_\perp)^2, \qquad (2.19)$$

with the square meaning two reflections at both the interfaces. Thus, for the transmission through N plates, we have:

$$P_\perp^{(N\,\text{plates})} = P_0(1 - R_\perp)^{2N}. \qquad (2.20)$$

Now, $R_\perp = \sin^2(\theta_B - \theta_t)$

$$\text{and } \theta_B + \theta_t = \pi/2 \Longrightarrow \theta_t = \frac{\pi}{2} - \theta_B,$$

$$\therefore \quad R_\perp = \sin^2(2\theta_B - \pi/2)$$

$$= \cos^2(2\theta_B) = \left(\cos^2\theta_B - \sin^2\theta_B\right)^2,$$

$$\because \quad \tan\theta_B = \frac{n_t}{n_i}.$$

$$R_\perp = \left(\frac{n_i^2 - n_t^2}{n_i^2 + n_t^2}\right)^2. \qquad (2.21)$$

Note that $n_i > n_t$ for internal reflection, while $n_t > n_i$ for external reflection. Both lead to the same expression for R (Eq. (2.21)). Substituting this expression into Eq. (2.20), we have:

$$P_{\perp}^{(N \text{ plate})} = P_0 \left[\frac{4 n_i^2 n_t^2}{(n_i^2 + n_t^2)^2} \right]^{2N}. \tag{2.22}$$

Let

$$D^{(N)} = \frac{P_{\parallel}}{P_{\perp}} (N \text{ plates}),$$

$$\therefore \quad D^{(N)} = \left[\frac{(n_i^2 + n_t^2)^2}{4 n_i^2 n_t^2} \right]^{2N},$$

$$D^{(N)} = \left[\frac{1}{4} \left(\frac{n_i}{n_t} + \frac{n_t}{n_i} \right)^2 \right]^{2N}. \tag{2.23}$$

Example 1: Glass/air, $n_i \cong 1$, $n_t = 1.5$.

$$N = 1, \ D^{(1)} = 1.38,$$
$$N = 10, \ D^{(10)} = 24.6,$$
$$N = 20, \ D^{(20)} = 627.5.$$

Example 2: Ge/air, $n_i \cong 1$, $n_t \cong 4$ for $10\,\mu$ radiation (CO_2 laser).

$$N = 1, \ D^{(1)} \cong 20.4,$$
$$N = 3, \ D^{(3)} \cong 8,500.$$

It is evident that the polarizer of this type is efficient when the material has a high index of refraction, such as Ge at $10\,\mu$. Thus, in manipulating CO_2 laser beams ($10{,}6\,\mu$), this type of Ge polarizer is often used because it is efficient and other types of inexpensive polarizers are not easily available.

(c) By setting the end windows (or ends, in the case of solids) of the active material inside a laser cavity at the Brewster angle, competition will favor the $\parallel$-component to the laser because it has no reflection loss in passing through the ends of the active medium. If the gain is not too high, the $\perp$-component will be suppressed completely so that the laser output is linearly polarized in the $\parallel$-direction.

2.3 Critical Angle and Total Internal Reflection

We now turn to the condition of $n_i > n_t$, i.e., the radiation passes from a denser medium to a less dense medium (see Fig. 2.8). All the consequences of the boundary conditions mentioned at the beginning of this chapter — $\theta_i = \theta_r$, the Snell's law, and k_i, k_r and k_t — lie in the same plane of incidence, and the Fresnel equations are still valid. In particular, we show in Fig. 2.9 the curves for $r_\perp$ and $r_\parallel$ as a function of θ_i. We can note the following two features:

(a) $r_\parallel$ passes through zero at $\theta_i = \theta_B'$.
(b) Both $r_\perp$ and $r_\parallel$ reach unity at $\theta_i = \theta_c$ and stay at unity for further increase of θ_i. θ_c. This is called the critical angle. At $\theta_i \geq \theta_c$, total internal reflection takes place.

θ_B' is again the Brewster angle. Using the same procedure as in the case of $n_i < n_t$, one can obtain (Note: This can be an exercise for the reader):

$$\tan \theta_B' = \frac{n_t}{n_i}, \quad (n_t < n_i) \tag{2.24}$$

and

$$\theta_B + \theta_B' = \pi/2. \tag{2.25}$$

The critical angle can be calculated through the condition that

$$r_\perp = r_\parallel = 1, \quad \text{at} \quad \theta_i = \theta_c.$$

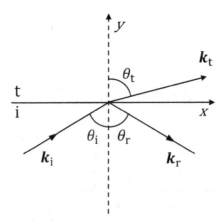

Fig. 2.8. Reflection and refraction at a boundary when the incident medium is denser than the transmission region.

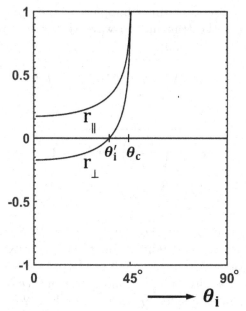

Fig. 2.9. $r_\parallel$ and $r_\perp$ as a function of the angle of incidence θ_i when the incident medium is denser.

Using either Eq. (2.3) or (2.4), one finds:

$$\theta_t = \pi/2, \quad \text{at} \quad \theta_i = \theta_c. \tag{2.26}$$

and Snell's law leads to:

$$\sin \theta_c = \frac{n_t}{n_i}. \tag{2.27}$$

It is interesting to note that when $\theta_i \geq \theta_c$ at which total internal reflection takes place, it does not mean that the electromagnetic field is now completely confined in the incident medium. In fact, the field penetrates into the transmission region exponentially. This can be seen as follows. Under the plane wave approximation, the transmitted field is:

$$E_t = E_{\text{ot}} \exp\{i(\boldsymbol{k}_t \cdot \boldsymbol{r} - \omega t + \epsilon_t)\},$$

where ϵ_t is a phase factor relative to the incident field. This phase factor depends only on n_i, n_t and θ_i, and can be obtained by calculating $t_\parallel$ using Eq. (2.6). Referring back to Fig. 2.8, which is defined as the

x–y plane and where everything happens in the plane of incidence, we can decompose the transmitted wave vector $\boldsymbol{k}_t$ into $(k_x\hat{\mathbf{i}} + k_y\hat{\mathbf{j}})$. Using Snell's law, we obtain:

$$E_t = E_{ot}\exp(-\alpha y)\exp\{i(\beta x - \omega t + \epsilon_t)\}, \qquad (2.28)$$

where

$$\alpha = k_t\sqrt{\left(\frac{n_i}{n_t}\right)^2\sin^2\theta_i - 1}, \qquad (2.29)$$

$$\beta = k_t\frac{n_i}{n_t}\sin\theta_i. \qquad (2.30)$$

Equation (2.28) shows that the transmitted field E_t penetrates the y-direction (transmission region) through the factor $\exp(-\alpha y)$. Moreover, it has a propagating component $\exp\{i(\beta x - \omega t + \epsilon_t)\}$ that propagates only in the x-direction, i.e., along the surface of the interface, but on the transmission side. E_t, as given by Eq. (2.29), is called the evanescent wave. It does not have any propagating component in the y-direction; hence, no transmitted ray goes into the transmitting medium.

Figure 2.10 shows several examples of total internal reflection.

2.4   Demonstration of Some Important Results of the Fresnel Equations

Some of the important features of the result of the Fresnel equations are:

(a) If $n_i < n_t$, $R \to 1$ as $\theta_i \to \pi/2$,
    i.e., the reflectance becomes unity at grazing incidence.
(b) If $n_i < n_t$, $R_\parallel = 0$ at $\theta_i = \theta_B$,
    i.e., the reflectance of the $\parallel$-component is zero at the Brewster angle.
(c) If $n_i > n_t$, $R = 1$ at $\theta_i \geq \theta_c$. That is, when the incident angle is equal to or greater than the critical angle, one has total internal reflection. However, there is an evanescent wave at the interface. It is emphasized because of its important applications in fiber optics, integrated optics, etc.

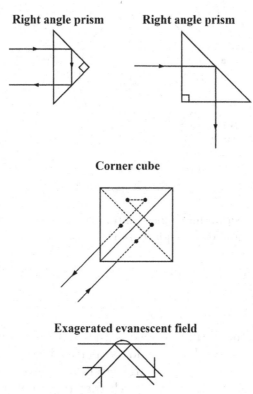

Fig. 2.10.   Examples of total internal reflection at work.

The following simple experiment gives a rather interesting demonstration of the three points above in front of a class of students.

Necessary equipment:

- One He-Ne laser, about 2 mW, preferably linearly polarized.
- A support for the laser.
- A small plastic bottle of alcohol (ethanol) with a squirting tip. (Note: Acetone, methanol, etc., would also do, but the vapor in a closed classroom will not be good for health.) A variation is to use a pipette or a glass tube.
- A hollow glass tube with an outer diameter of less than $\frac{1}{2}$ cm.
- A right angle prism, at least 2.5 cm x 2.5 cm x 2.5 cm.
- A lab-jack or any appropriate support for the prism so that the height can be adjusted.
- A small shallow dish to collect the alcohol.

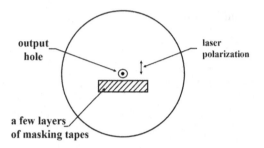

Front view of laser head

Fig. 2.11.   Schematic set up for a demonstration.

- Any piece of rectangular glass plate, e.g., a microscopic substrate. However, the longer the plate is the easier it is to perform the demonstration.
- A piece of white paper about the size of a page of this book (or larger).

Preparation

Draw a line on the output surface of the laser head and just below the output hole. This line should be perpendicular to the laser's polarization direction. Tape several layers of masking tapes on the output surface of the laser head such that the tape's edge lines up with the line just drawn (see Fig. 2.11). Rotate the laser around its axis so that the polarization is almost vertical (or the tapes are almost horizontal). Fix the laser on the support. Put everything on a cart and wheel them to the classroom. The demonstration can be done on the cart or a classroom table.

Demonstration procedure

Warning: Never point the laser beam towards anybody's eyes nor should one ever look into a laser beam. The following order is not important, but I happen to like it personally.

(1) With the demonstrator facing the class and the cart in front of him, the laser is pointed towards one of the side walls (left or right, depending on convenience). Turn on the laser and ask the class to look at the red laser spot on the wall. It is important that the height of the laser beam be set above the head level of the class to avoid accidentally pointing the laser beam into anyone's eye.

(2) Put the glass plate into the laser beam so that its surface is nearly perpendicular to the beam. Make a remark that the intensity of the spot on the wall is almost not altered because the reflectance from the glass surfaces is about 8%.

(3) Now quickly increase the angle of incidence such that the reflected light beam also appears on the side wall. To have a steady hold of the plate, touch the shorter side of the plate on the output surface of the laser head. Note that the transmitted beam's intensity is still very strong and appears unaffected.

(4) Now, carefully and slowly, increase the angle of incidence until the reflected spot is very near the transmitted spot. Note that the transmitted beam weakens rather quickly while the reflected beam increases in intensity. If the demonstrator is careful, he should be able to make the transmitted beam almost disappear. This is a remarkable phenomenon because the transmitted beam, initially almost unaffected, can be reduced significantly.

(5) Now hold the glass plate with its shorter side supported by the horizontal edge of the layers of masking tape on the output surface of the laser head (Fig. 2.12). Tell the class that the glass plate will be turned with the pivot at the edge of the masking tape, and then explain why the reflected light is $R_\parallel$ with such an orientation.

(6) Start with the plate at almost grazing incidence and ask the class to watch the movement of the reflected spot. Gradually decrease the angle of incidence and the reflected spot will rise and scan across the ceiling. When the angle of incidence approaches the Brewster angle, the intensity of the laser spot on the ceiling decreases, and at $\theta_i = \theta_B$, there is no reflection. Further reduction of the angle of incidence makes the reflected beam reappear on the ceiling. Again, it is remarkable that the reflected beam, while very strong at grazing incidence, is zero at the Brewster angle.

(7) Put away the glass plate, set the prism inside the shallow disk, and put them on the lab-jack.

(8) Adjust the lab-jack so that the prism intercepts the laser beam. Let the laser beam enter from one side of the right-angled surface and be totally internally reflected at the hypotenuse surface. Turn the cart slightly such that the hypotenuse surface of the prism faces the class.

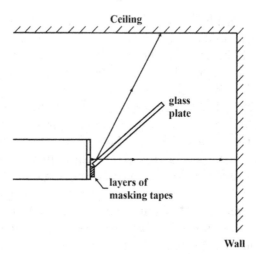

Fig. 2.12.   Another view of the schematic set up for the same demonstration.

(9) Turn off the light in the room. Note a small spot of scattering light on the hypotenuse surface where the laser beam is totally reflected internally (see Fig. 2.13). However, holding the white paper in front of this spot will not show any transmitted beam of the laser in any direction. The scattered light shows that the light penetrates across the interface and is scattered by the microscopic irregularities of the surface or dust.

(10) Notice that the evanescent field has a penetration term $e^{-\alpha y}$ in the direction normal to and outward from the prism surface (Fig. 2.13), and a propagation term $\exp[i(\omega t - \mathrm{k}x)]$ propagating along the surface in the x-direction. Hold the glass tube vertically and touch the prism surface where the light is totally reflected. Immediately, there is a horizontal "fan" of light coming out of the glass tube (see Fig. 2.14). This is because the evanescent field penetrates the tube and the propagation component "guides" the beam around the tube through multiple reflections and transmissions, resulting in a horizontal fan of light.[1,2] This shows that the evanescent field has penetrated across the interface. (See **Section 2.5** for more discussion.)

---

[1]S. L. Chin, "Optical fan levelling system", Canadian Patent No. 9555393, October 1974.

[2]S. L. Chin and K. A. Mace, "Optical fan levelling system", US Patent No. 3984154, October 1975.

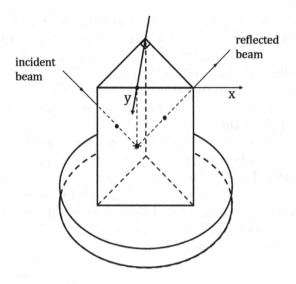

Evanescent field $\sim \mathbf{e}^{-\alpha y}\,\mathbf{e}^{-i(\omega t - kx)}$

Fig. 2.13.   A demonstration of the total internal reflection and evanescent wave.

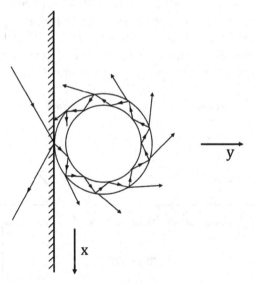

Fig. 2.14.   Frustrated total internal reflection.

(11) Put away the glass tube. Hold the white page of paper at the side of the prism (on the downstream side of the x-direction, see Fig. 2.13). Squirt some alcohol on the surface where the light is totally reflected. A bright splash of light coming from the evanescent field propagating along the surface hits the paper. This is another demonstration of the penetration and propagation nature of the evanescent field.

(12) A bit of light show: Touch the wet tip of the alcohol squirter at the spot where there is total internal reflection and observe the reflected beam (Fig. 2.13) on a screen (or the white page of paper). Beautiful interference fringes of different forms appear and disappear dynamically. Interference of the evanescent wave in the thin alcohol film on the prism surface and the evaporation process cause all these.

2.5 Making Good Use of the Evanescent Field

The evanescent field (Eqs. (2.28) and (2.29)) has been put into good use in optoelectronics and photonics. One popular application is to couple laser (optical) beams in and out of thin films for integrated optics, etc. This is shown in Fig. 2.15. By pressing the prisms onto the thin film, the air space between the prism and thin film is reduced,

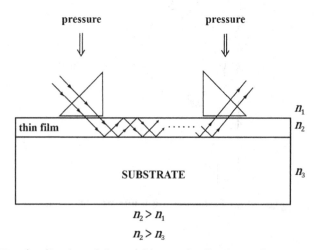

Fig. 2.15. Application of the total internal reflection and evanescent wave.

allowing the evanescent field to penetrate the film and propagate inside the film again by total internal reflection ($\because n_2 > n_1, n_2 > n_3$).

This possibility is already shown in Section 2.4, demonstration procedure #10. Many other uses are applied to different situations but the principle of <u>frustrating the total internal reflection, which allows the evanescent field to go through</u>, is the same. For instance, one wants to obtain some light out of an optical fiber through which a certain light wave propagates. Since light propagation through fibers is essentially by total internal reflection (a more rigorous description by solving the Maxwell equations can be found in any fiber optics book), touching the bare fiber surface with another transparent material of an appropriate index will induce some leakage through the frustration of the total internal reflection in the fiber.

Chapter 3

Resonator: A Geometrical View

3.1 Introduction

In this chapter, we will discuss a resonant cavity that is one of the basic optical elements for laser operation by confining light in a space, e.g., between two reflection mirrors, to be amplified. Simply speaking, the laser beam inside the cavity will automatically bounce back and forth between the two mirrors. This can be understood by referring to Fig. 3.1, which shows a beam bouncing back and forth between two parallel plane mirrors. In reality, this is not a rule, but a special situation, in which the beam is perpendicular to the surfaces of the two mirrors. Any other inclination of the light beam with respect to the mirror surfaces will lead to a walk-off and become a total loss, i.e., no laser oscillation is possible, as shown in Fig. 3.2. Of course, one immediately thinks of using a better system, namely, a pair of concave mirrors. This is indeed true, but it does not mean that any pair of concave mirrors will allow a stable laser oscillation. Without a proper choice of the radii of curvature of the mirrors and the distance between them, the light beam in the cavity will still walk-off. We shall study in this chapter the condition under which a pair of mirrors will allow a stable laser oscillation. The approach is based on geometrical optics because it is much easier to understand. More rigorous diffraction approaches can be found in more advanced laser textbooks.

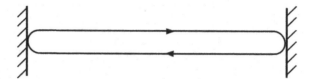

Fig. 3.1. A cavity bounded by two plane parallel mirrors.

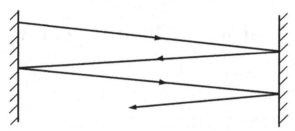

Fig. 3.2. A light beam walks off the cavity bounded by a pair of plane parallel mirrors.

3.2   General Considerations

We look at the most general laser cavity bounded by two concave mirrors whose radii of curvature can be any positive or negative values, as shown in Fig. 3.3. It is well-known in geometrical optics that the focal length of a concave mirror of radius of curvature R is $f = R/2$. Hence, the effect of such a mirror on a beam of light is equivalent to that of a lens with focal length $f = R/2$. As such, we find out similarities in Figs. 3.3 and 3.4. In Fig. 3.3, a beam of light is assumed to start at point P just in front of R_2 towards R_1 and be reflected to point R just in front of R_1. This is equivalent to passing the beam of light through a lens of focal length $f_1 = R_1/2$, as shown in Fig. 3.4. From point R in Fig. 3.3, the beam keeps on propagating towards R_2 and is reflected to point T just in front of R_2. This is equivalent to the continuation of the ray PQR towards S and T in Fig. 3.4, passing the equivalent lens of focal length $f_2 = R_2/2$. This operation completes a round trip of the beam between the two mirrors. For an oscillation to take place, the beam in the cavity, in principle, will bounce back and forth an infinite number of times. This is equivalent to having a beam of light pass through an infinite pair of equivalent lenses in series, as shown in Fig. 3.5. The condition under which there exists such an oscillation is that the beam does

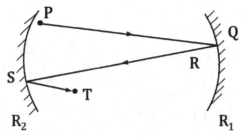

Fig. 3.3. A cavity bounded by two spherical mirrors.

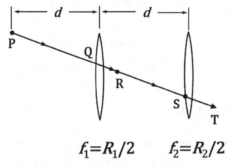

$$f_1 = R_1/2 \qquad f_2 = R_2/2$$

Fig. 3.4. Propagation through a pair of lenses, which is equivalent to the round trip reflection between two spherical mirrors.

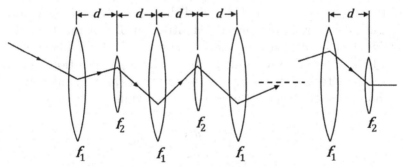

Fig. 3.5. Multiple reflections between two spherical mirrors are equivalent to the transmission through many pairs of lenses.

not walk-off the mirrors, i.e., the beam is forever bouncing within the bounds of the mirrors. This is equivalent to the beam in Fig. 3.5 always coming back and passing through the lens series to infinity.

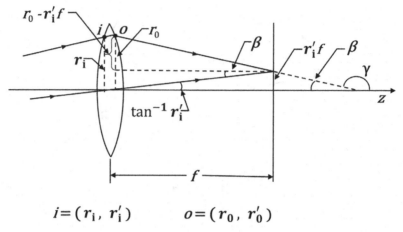

$$i = (r_i, r'_i) \qquad o = (r_0, r'_0)$$

Fig. 3.6. Transmission through a lens.

3.3 Case of One Lens

We shall now make use of geometrical optics in the matrix form to analyze Fig. 3.5. We start by analyzing the case of a beam passing through a single lens. The analysis is based on two assumptions: a *paraxial ray* and a very thin lens. A *paraxial ray* means that the beam is confined to a narrow region around the axis of the lens. It means that the angle between the beam and the axis of the lens is small.

Figure 3.6 shows an exaggerated situation of a parallel beam of light incident on a thin lens and exit toward the focal plane where the light focuses. The incident beam at the input position i of the lens is characterized by r_i and r'_i, where r_i is the vertical distance from point i to the axis of the lens (z-axis) while

$$r'_i \equiv \frac{dr_i}{dz}$$

is the slope of the ray at r_i with respect to the z-axis. Similarly, at the output side of the lens, the output beam at point 0 is characterized by r_0 and r'_0, which are similarly defined. We shall find the relationship between (r_i, r'_i) and (r_0, r'_0). It is obvious that

$$r_0 = r_i \tag{3.1}$$

because of the thin lens approximation. Inspecting Fig. 3.6, we see that

$$r_o' \equiv \frac{dr_o}{dz} = \tan\gamma = \tan(\pi - \beta),$$

$$= -\tan\beta = -\frac{r_o - r_i' f}{f}. \tag{3.2}$$

Rewriting Eqs. (3.1) and (3.2):

$$r_o = r_i + o \cdot r_i',$$

$$r_o' = \left(-\frac{1}{f}\right) r_i + r_i',$$

i.e., $$\begin{bmatrix} r_o \\ r_o' \end{bmatrix} = \begin{bmatrix} 1 & 0 \\ -\dfrac{1}{f} & 1 \end{bmatrix} \begin{bmatrix} r_i \\ r_i' \end{bmatrix}. \tag{3.3}$$

Equation (3.3) is the matrix representation of a paraxial ray passing through a thin lens, showing the relationship of the ray at the two sides (i and 0) of the lens. We need also the matrix representation of the propagation in free space.

Figure 3.7 shows a ray passing through two positions z_1 and z_2 in space. The "coordinates" of the ray at z_1 and z_2 are (r_i, r_i') and

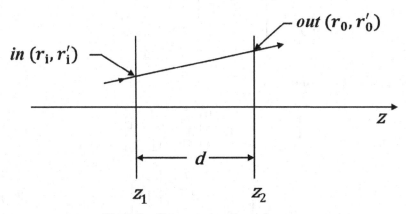

Fig. 3.7. Transmission through space.

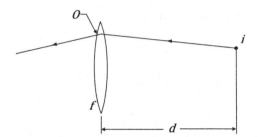

Fig. 3.8. Transmission through space and a lens.

(r_o, r'_o). We immediately see that

$$r_o = r_i + r'_i d,$$
$$r'_o = r'_i.$$

In matrix form, they become:

$$\begin{bmatrix} r_o \\ r'_o \end{bmatrix} = \begin{bmatrix} 1 & d \\ 0 & 1 \end{bmatrix} \begin{bmatrix} r_i \\ r'_i \end{bmatrix}. \tag{3.4}$$

Thus, combining the above two cases, we have the situation shown in Fig. 3.8, in which we have reversed the propagation direction for the sake of understanding the solution. The "coordinates" at 0 are thus:

$$\begin{bmatrix} r_o \\ r'_o \end{bmatrix} = \begin{bmatrix} lens \\ matrix \end{bmatrix} \begin{bmatrix} free \\ space \\ matrix \end{bmatrix} \begin{bmatrix} r_i \\ r'_i \end{bmatrix}.$$

Using Eqs. (3.3) and (3.4):

$$\begin{bmatrix} r_o \\ r'_o \end{bmatrix} = \begin{bmatrix} 1 & 0 \\ -\dfrac{1}{f} & 1 \end{bmatrix} \begin{bmatrix} 1 & d \\ 0 & 1 \end{bmatrix} \begin{bmatrix} r_i \\ r'_i \end{bmatrix},$$

$$= \begin{bmatrix} 1 & d \\ -\dfrac{1}{f} & -\dfrac{d}{f} + 1 \end{bmatrix} \begin{bmatrix} r_i \\ r'_i \end{bmatrix}. \tag{3.5}$$

3.4 Case of Two Lenses and Equivalence to One Round Trip in the Cavity

As mentioned in Section 3.2, having a beam of light making one round trip between the mirrors of the laser cavity is equivalent to passing the beam through two equivalent lenses, as shown in Fig. 3.9. The light beam starts again from the right at point i and exits the second lens at point 0. The "coordinates" of the beam at 0 are thus:

$$
\begin{bmatrix} r_o \\ r'_o \end{bmatrix} = \begin{bmatrix} lens\ f_2 \end{bmatrix} \begin{bmatrix} free \\ space \\ d \end{bmatrix} \begin{bmatrix} lens \\ f_1 \end{bmatrix} \begin{bmatrix} free \\ space \\ d \end{bmatrix} \begin{bmatrix} r_i \\ r'_i \end{bmatrix}.
$$

It is now straightforward to calculate the resultant matrix using Eqs. (3.3) to (3.5). The detailed calculation is left as an <u>exercise</u> for the readers, and the result is:

$$
\begin{bmatrix} r_o \\ r'_o \end{bmatrix} = \begin{bmatrix} A & B \\ C & D \end{bmatrix} \begin{bmatrix} r_i \\ r'_i \end{bmatrix}, \tag{3.6}
$$

where

$$
A = 1 - \frac{d}{f_1}, \tag{3.7}
$$

$$
B = d \left(2 - \frac{d}{f_1} \right), \tag{3.8}
$$

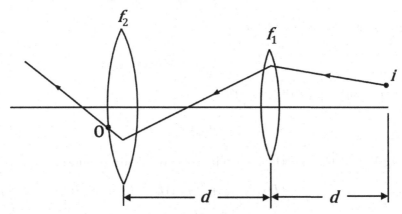

Fig. 3.9. Round trip reflection between two spherical mirrors is equivalent to the transmission through space and two lenses.

$$C = -\frac{1}{f_1} - \frac{1}{f_2}\left(1 - \frac{d}{f_1}\right), \tag{3.9}$$

$$D = -\frac{d}{f_2} + \left(1 - \frac{d}{f_2}\right)\left(1 - \frac{d}{f_1}\right). \tag{3.10}$$

3.5 General Case of a Biperiodic Lens Series and the Condition for a Stable Resonator

A biperiodic lens series shown in Fig. 3.5, which is equivalent to laser oscillation in a cavity between two mirrors (Section 3.2), can now be considered. Each unit section of the series is a pair of lenses shown in Section 3.4. We consider a general section, say the sth section. From Eq. (3.6), we have:

$$\begin{bmatrix} r_{s+1} \\ r'_{s+1} \end{bmatrix} = \begin{bmatrix} A & B \\ C & D \end{bmatrix} \begin{bmatrix} r_s \\ r'_s \end{bmatrix}, \tag{3.11}$$

where $i \to s$ and $0 \to s+1$, and A, B, C and D are given by Eqs. (3.7) to (3.10).

$$\text{i.e.,} \quad r_{s+1} = Ar_s + Br'_s, \tag{3.12}$$

$$r'_{s+1} = Cr_s + Dr'_s, \tag{3.13}$$

$$\text{Eq. (3.12)} \to \quad r'_s = \frac{1}{B}(r_{s+1} - Ar_s). \tag{3.14}$$

Increasing s by unity, we have:

$$r'_{s+1} = \frac{1}{B}(r_{s+2} - Ar_{s+1}). \tag{3.15}$$

Substituting Eq. (3.13) into it,

$$Cr_s + Dr'_s = \frac{1}{B}(r_{s+2} - Ar_s + 1). \tag{3.16}$$

Using Eq. (3.14) and then simplifying, Eq. (3.16) becomes:

$$r_{s+2} - (D + A)r_s + 1 + (AD - BC)r_s = 0. \tag{3.17}$$

But

$$\begin{vmatrix} A & B \\ C & D \end{vmatrix} = 1,$$

as can be verified using Eqs. (3.7) to (3.10). Thus, Eq. (3.17) simplifies to:

$$r_s + 2 - 2br_s + 1 + r_s = 0, \tag{3.18}$$

where

$$b \equiv \frac{1}{2}(A + D). \tag{3.19}$$

We try the following solution for Eq. (3.18):

$$r_s = r_0 e^{is\theta}. \tag{3.20}$$

By substituting into Eq. (3.18), we have, after some slight simplification:

$$e^{i2\theta} - 2be^{i\theta} + 1 = 0, \tag{3.21}$$

$$\text{i.e., } e^{i\theta} = b \pm i\sqrt{1 - b^2}. \tag{3.22}$$

Similarly, $e^{-is\theta}$ is also a solution of Eq. (3.18). The general solution of Eq. (3.18) is thus a linear combination of $e^{-is\theta}$ and $e^{is\theta}$, which can be represented by a sine function.

Thus,

$$r_s = r_{\max} \sin(s\theta + \delta), \tag{3.23}$$

where $r_{\max}$ and δ are some appropriate real quantities. The condition of stable laser oscillation means that the light beam will always pass through the biperiodic lens series (Section 3.2). That is to say, Eq. (3.23) should always oscillate between two appropriate values of $(r_{\max})$ and $(-r_{\max})$. This demands that θ be real. From Eq. (3.22), this means:

$$|b| \leq 1,$$

$$\text{i.e., } \left| \frac{1}{2}(A + D) \right| \leq 1, \quad (\text{from Eq. (3.19)})$$

or

$$\left| 1 - \frac{d}{f_1} - \frac{d}{f_2} + \frac{d^2}{2f_1 f_2} \right| \leq 1. \tag{3.24}$$

After some simplification, which we leave as an <u>exercise</u> for the readers, we obtain:

$$0 \leq \left(1 - \frac{d}{2f_1} \right) \left(1 - \frac{d}{2f_2} \right) \leq 1. \tag{3.25}$$

Definition: g parameter

$$g \equiv 1 - \frac{d}{2f}, \quad \text{for a lens.} \tag{3.26}$$

Since $f = R/2$,

$$g \equiv 1 - \frac{d}{R}, \quad \text{for a mirror.} \tag{3.27}$$

Equation (3.25) becomes:

$$0 \leq g_1 g_2 \leq 1, \tag{3.28}$$

which is thus the condition for a stable resonator where

$$g_1 = 1 - \frac{d}{R_1} \tag{3.29}$$

and

$$g_2 = 1 - \frac{d}{R_2}. \tag{3.30}$$

The condition (3.28) is shown graphically in Fig. 3.10. The shaded regions are regions of stable resonators, including the origin and two

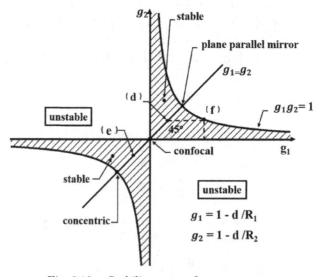

Fig. 3.10.   Stability curves for a resonator.

axes where $g_1 g_2 = 0$, and also any point on the hyperbolas where $g_1 g_2 = 1$. Note: The signs of R_1 and R_2 are such that they are positive if the inner surfaces are concave as shown in Fig. 3.3. Otherwise, they are negative.

Examples:

(a) Plane parallel mirrors

$$R_1 = R_2 = \infty$$

$$g_1 = g_2 = 1$$

$$g_1 g_2 = 1 \quad \Rightarrow \quad \text{stable}$$

(b) Confocal resonator

$$R_1 = R_2 = d$$

$$g_1 = g_2 = 0 \quad \Rightarrow \quad \text{stable}$$

(c) Concentric resonator

$$R_1 = R_2 = d/2$$

$$g_1 = g_2 = -1$$

$$g_1 g_2 = 1 \quad \Rightarrow \quad \text{stable}$$

(d)

$$R_1 = R_2 = 2d$$

$$g_1 = g_2 = 0.5$$

$$g_1 g_2 = 0.25 \quad \Rightarrow \quad \text{stable}$$

(e)

$$R_1 = R_2 = \tfrac{2}{3}d$$

$$g_1 = g_2 = -0.5$$

$$g_1 g_2 = 0.25 \quad \Rightarrow \quad \text{stable}$$

(f)

$$R_1 = -d, \quad d \le R_2 \le 2d$$

$$g_1 = 2, g_2 \le \frac{1}{2}$$

$$g_1 g_2 \le 1 \quad \Rightarrow \quad \text{stable}$$

This last example shows that one can have a stable resonator even if one of the mirrors is convex, so long as the other mirror compensates for the negative effect (diverging) of the convex surface.

(g) $$R_1 > +d, \quad 0 < R_2 < +d$$

Hence,

$$\frac{d}{R_1} < 1 \quad \text{and} \quad \frac{d}{R_2} > 1.$$

$$g_1 = 1 - \frac{d}{R_1} > 0 \quad \text{and} \quad g_2 = 1 - \frac{d}{R_2} < 0.$$

Result : $g_1 g_2 < 0$, i.e., unstable.

Note: Even if R_1 and R_2 are concave, it is not a guarantee that the resonator will be stable.

Chapter 4

The Laser

4.1 Definition of a Laser Oscillator

The *laser* (light amplification by stimulated emission of radiation) is essentially a resonator that contains a light amplifier whose amplification process is stimulated emission (Fig. 4.1). Such a combination becomes an oscillator when it is sustained to operate in equilibrium, giving out a beam of light from one or more ends of the resonator. The condition under which the oscillator operates in equilibrium is:

$$\text{gain} = \text{loss}.$$

More precisely, let us consider Fig. 4.2. Assume that a pulse of a light beam of energy E_0 starts at $t = 0$, $z = 0$, and propagates through the amplifying medium. When it reaches the mirror at the right end, it is partially reflected back into the amplifier, reaching the mirror at the left end, and is again partially reflected back towards the starting point $z = 0$, thus making a round trip. If the laser operation is to be sustained in equilibrium, the net gain and net loss during any round trip cycle should be equal. In other words, the energy E in the light pulse remains unchanged after any round trip.

$$\text{i.e.,} \quad E(\text{round trip}) = E_0. \tag{4.1}$$

But

$$E(\text{round trip}) = E_0 e^{g \cdot 2l_\text{m}} e^{-\alpha_\text{C} \cdot 2l}, \tag{4.2}$$

where g is the gain coefficient (cm^{-1}) of the amplifying medium of length l_m, and α_C is the total loss coefficient (cm^{-1}) of the oscillator. No saturation is assumed. The factor 2 in the exponentials

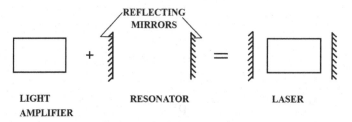

Fig. 4.1.   Schematic definition of a laser oscillator.

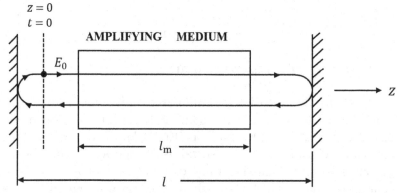

Fig. 4.2.   Schematic illustration of a light beam making a round trip in a cavity.

means "round trip". Here, we have assumed exponential gain and exponential loss, the latter being a description of most decay processes governed by Beer's law (i.e., exponential decay). They also imply that the energy gain (loss) per unit length is proportional to the energy,

i.e.,
$$\pm \frac{\mathrm{d}E}{\mathrm{d}z} \propto E$$

or

$$\binom{+}{-} \frac{\mathrm{d}E}{\mathrm{d}z} = \binom{g}{\alpha_{\mathrm{C}}} E,$$

$$E(z) = \begin{cases} E_0 e^{gz}, & \text{gain} \\ E_0 e^{-\alpha_{\mathrm{C}} z} & \text{loss} \end{cases}.$$

Going back to Eqs. (4.1) and (4.2), we have:

$$E_0 e^{g2l_\mathrm{m}} e^{-\alpha_C 2l} = E_0,$$

or

$$2(gl_\mathrm{m} - \alpha_C l) = 0,$$

$$gl_\mathrm{m} = \alpha_C l, \quad \text{(threshold condition)}. \tag{4.3}$$

Discussion:

(1) *Loss* α_C

The coefficient α_C is essentially a lumped factor representing loss through

(a) transmission of the two mirrors,
(b) scattering, absorption and diffraction at the two mirrors,
(c) absorption, scattering and diffraction in the amplifying medium, and
(d) reflection, scattering, absorption and diffraction at the two ends of the amplifying medium.

In principle, these losses can be measured experimentally, either separately or together. We take this opportunity to link the loss α_C with the quality factor Q, which is commonly used in electrical engineering. In a passive resonator (no gain), Q is defined as:

$$Q \equiv \omega \frac{E}{\frac{-dE}{dt}},$$

where E is the radiation energy stored in the resonator at time t, $-\frac{dE}{dt}$ is the average energy dissipation per second, and ω is the angular frequency of the radiation, $\omega = 2\pi\nu$ (ν is the frequency in Hertz). Thus

$$\frac{dE}{dt} = -\frac{\omega}{Q} E,$$

and

$$E = E_0 e^{-\frac{\omega}{Q} t}, \tag{4.4}$$

where E_0 is the energy in the cavity at $t = 0$. Equation (4.4) illustrates Beer's law and the role of Q in the law. Assume now that we inject a radiation pulse of energy E_0 into the empty cavity at $t = 0$

and let it bounce back and forth between the mirrors N times. The energy in the pulse after N round trips (in t seconds) is:

$$E = E_0 e^{-N \cdot 2l \cdot \alpha_C},$$

where

$$N \equiv \frac{t}{2l/c},$$

$$\frac{2l}{c} = \text{one round trip time},$$

and c is the speed of light in the cavity.

Hence,

$$E = E_0 e^{-\dfrac{t}{2l/c} \cdot 2l \cdot \alpha_c},$$

$$= E_0 e^{-c\alpha_c t}. \tag{4.4$'$}$$

Comparing with Eq. (4.4), we have

$$\frac{\omega}{Q} = c\alpha_C. \tag{4.4$''$}$$

From either Eq. (4.4) or (4.4$''$), we see that the higher the value of Q, the lower is the loss, and vice versa. Thus, changing Q means changing the loss; hence, the name Q-switching for some lasers (Chapter 8).

Thus, Eq. (4.3) becomes:

$$Q = \frac{\omega l}{g l_m C} = \frac{2\pi l}{g l_m \lambda}, \text{(threshold condition)} \tag{4.4$'''$}$$

where

$$\lambda \equiv \text{wavelength},$$

$$= \frac{c}{\nu}.$$

(2) *Gain g*

The gain coefficient g is due to stimulated emission between two energy levels of the active medium. Because emission is always accompanied by the possibility of absorption, one has to consider the two together. The following section discusses such phenomena.

4.2 Stimulated Emission

Stimulated emission is a general phenomenon in nature. It is the result of the interaction of an electromagnetic wave with matter. Whenever there is an electromagnetic wave of an appropriate frequency interacting with a material with some appropriate energy levels, there will be stimulated emission. Thus, one can say that any material might become a laser material (i.e., amplifying medium) under some appropriate conditions. However, the reality is much more complicated. Because there are many more loss mechanisms than gain (stimulated emission), one has to prepare the medium carefully in order that the gain can overcome the loss. We shall consider first of all the simplest system in thermal equilibrium so as to illustrate the relationship between emission and absorption. This analysis was first published by Einstein in 1917 (Einstein, 1917).

Figure 4.3 shows such a simple system, a 2-level system, labeled 1 and 2. The energy of level $i(i = 1, 2)$ with respect to an arbitrary reference is E_i. The two levels could be any pair of energy levels (electronic, vibrational or rotational levels) of a material that can be coupled by an electromagnetic field of an appropriate frequency ν such that

$$h\nu = E_2 - E_1, \tag{4.5}$$

where h is the Planck constant.

When such a purely 2-level system (which we now call an atom) has an energy E_1 with respect to an arbitrary reference, we say that

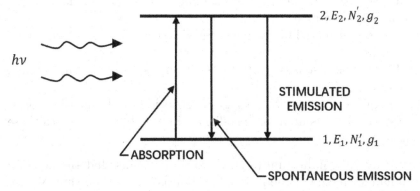

Fig. 4.3. A 2-level system showing absorption and emissions of radiation.

the atom is in level 1. If it has an energy E_2, it is in level 2. When it is in level 1, it can absorb radiation of frequency ν that satisfies Eq. (4.5). When it is in level 2, it can emit radiation through either spontaneous emission or stimulated emission. The stimulated emission is a function of the radiation density that is present in the region where the atom is situated. The emitted radiation has a frequency ν. Even when there is no radiation present in the region, the atom will emit a photon $h\nu$ by itself. Analysis by quantum mechanics shows that this is a purely quantum effect that can be viewed essentially as the interaction of the random vacuum fluctuation with the atom "stimulating" the latter to emit a photon "randomly" if it is already excited into level 2. The statistical nature of this "random" spontaneous emission will be seen immediately in what follows. We note here that all the above statements can be rigorously calculated by quantum mechanics. Interested readers can consult more advanced books on quantum electrodynamics or quantum optics.

We make the following assumptions:

(a) Each of the levels is degenerate with a *multiplicity* $g_i(i = 1, 2)$. Definition: *multiplicity* $g_i(i = 1, 2) \equiv$ the number of states with the same energy $E_i(i = 1, 2)$. Each state in level i has an equal number of atoms N_i'.

(b) There are a total of N_{tot} 2-level atoms each having a discrete energy $E_i(i = 1, 2)$.
Thus,

$$g_1 N_1' + g_2 N_2' = N_{\text{tot}}. \tag{4.6}$$

(c) These atoms bathe in an electromagnetic radiation field within an isolated region.

(d) The radiation field is monochromatic with frequency ν satisfying Eq. (4.5).

(e) The whole isolated (atoms + field) system is in thermal equilibrium.

(f) Energy exchange is through absorption and emission only. Hence, there is no atomic collision. Figure 4.3 summarizes these assumptions.

Now, thermal equilibrium (assumption (e)) means that the net rate of dynamic change of the isolated system is zero. (Note that Nature is always in a dynamic state, i.e., everything in the microscopic scale is in motion unless the temperature becomes $T = 0\,K$, which itself is defined as a state of matter whose microscopic constituents are at rest.) That is, if we consider only the number of atoms N_1 in level 1, the net rate of change of $N_1 = 0$, or

$$\left(\frac{\partial N_1}{\partial t}\right)_{\text{net}} = 0, \tag{4.7}$$

where

$$N_i \equiv g_i N_i', \quad i = 1, 2. \tag{4.8}$$

The variation in N_1 is caused by the absorption of photons from level 1 into level 2 (decrease of N_1) and the emission of radiation from level 2 back to level 1 (decrease of N_2 leading to the increase of N_1) by assumption (f).

$$\text{i.e.,} \quad 0 = \left(\frac{\partial N_1}{\partial t}\right)_{\text{net}} = -\left(\frac{\partial N_1}{\partial t}\right)_{\text{abs.}}$$
$$+ \left(\frac{\partial N_2}{\partial t}\right)_{\text{spon.em.}} + \left(\frac{\partial N_2}{\partial t}\right)_{\text{st.em.}}. \tag{4.9}$$

In Eq. (4.9), the negative sign means "decrease" while the positive sign means "increase". The abbreviations "abs.", "spon.em." and "st.em." mean "absorption", "spontaneous emission" and "stimulated emission", respectively. Now, common sense suggests to us the following linear relationship:

$$\left(\frac{\partial N_1}{\partial t}\right)_{\text{abs.}} \propto \rho(\nu) g_1 N_1',$$

$$\left(\frac{\partial N_2}{\partial t}\right)_{\text{st.em.}} \propto \rho(\nu) g_2 N_2',$$

$$\left(\frac{\partial N_2}{\partial t}\right)_{\text{spon.em.}} \propto g_2 N_2', \text{(independent or $\rho(\nu)$)},$$

where $\rho(\nu)$ is the radiation density per unit frequency at frequency ν in the interaction region, i.e., $[\rho(\nu)] = J/\{(\text{m}^3).\text{Hz}\}$. Here, only the absolute values are considered. Positive and negative signs are taken care of in Eq. (4.9),

$$\text{i.e.,} \quad \left(\frac{\partial N_1}{\partial t}\right)_{\text{abs.}} = B_{12}\rho(\nu)g_1 N_1', \tag{4.10}$$

$$\left(\frac{\partial N_2}{\partial t}\right)_{\text{st.em.}} = B_{21}\rho(\nu)g_2 N_2', \tag{4.11}$$

$$\left(\frac{\partial N_2}{\partial t}\right)_{\text{spon.em.}} = A_{21}g_2 N_2', \tag{4.12}$$

where A_{21}, B_{12} and B_{21} are the constants of proportionality. These linear relationships are indeed valid in a more rigorous quantum mechanical analysis so long as the radiation density is not high. Substituting Eqs. (4.10) to (4.12) into (4.9),

$$-B_{12}\rho(\nu)g_1 N_1' + B_{21}\rho(\nu)g_2 N_2' + A_{21}g_2 N_2' = 0$$

or

$$\frac{g_2 N_2'}{g_1 N_1'} \equiv \frac{N_2}{N_1} = \frac{B_{12}\rho(\nu)}{B_{21}\rho(\nu) + A_{21}}. \tag{4.13}$$

But from thermodynamics, in thermal equilibrium,

$$\frac{N_2}{N_1} = \frac{g_2}{g_1}\exp\left[-\frac{E_2 - E_1}{kT}\right], \tag{4.14}$$

where T is the temperature, and k is the Boltzmann constant. Equating (4.13) and (4.14) and solving for $\rho(\nu)$ gives:

$$\rho(\nu) = \frac{(A_{21}/B_{21})}{\left(\frac{g_1}{g_2}\right)\left(\frac{B_{12}}{B_{21}}\right)\exp\left(\frac{h\nu}{kT}\right) - 1}, \tag{4.15}$$

where $h\nu = E_2 - E_1$ from Eq. (4.5).

A radiation field in thermal equilibrium with an atomic system means that it is a radiation field from a black body. Such a field has a Boltzmann distribution given by Boltzmann's law:

$$\rho(\nu) = \frac{8\pi\nu^2}{c^3}\frac{h\nu}{(e^{h\nu/kT}) - 1}. \tag{4.16}$$

Comparing Eqs. (4.15) and (4.16), we obtain:

$$\frac{A_{21}}{B_{21}} = \frac{8\pi\nu^2}{c^3}h\nu, \tag{4.17}$$

$$B_{12} = \frac{g_2}{g_1}B_{21}. \tag{4.18}$$

Discussion:

(1) A_{21}, B_{12} and B_{21} are called the *Einstein A, B coefficients*.

(2) If $g_2 = g_1 = 1$ (i.e., no degeneracy), Eq. (4.18) gives:

$$B_{12} = B_{21}. \tag{4.18'}$$

From Eqs. (4.10) and (4.11), we see that $B_{12}\rho(\nu)$ and $B_{21}\rho(\nu)$ are the probability per second of absorption and stimulated emission, respectively. Therefore, if there is no degeneracy, i.e., if we have only two discrete levels,

$$B_{12}\rho(\nu) = B_{21}\rho(\nu), \quad \text{(from Eq. (4.18'))}$$

or *the probabilities per second of absorption and stimulated emission are equal*. This is a very important result. It means that, physically, because there is also the contribution of spontaneous emission, the total down transition probability per second ($2 \to 1$) is always greater than the up transition ($1 \to 2$) probability per second. Since thermal equilibrium requires the total rate of change of N_1 and N_2 to be equal, N_1 will always be greater than N_2 unless $T = \infty$ (see below).

(3) Pure spontaneous emission can be described using Eq. (4.12) as:

$$-\frac{\partial N_2}{\partial t} = A_{21}N_2,$$

assuming $\rho(\nu) = 0$ and there are N_{20} atoms excited to level 2 at $t = 0$. The solution of this equation is:

$$N_2 = N_{20}\exp(-A_{21}t)$$

or

$$N_2 = N_{20}\exp\left(-\frac{t}{\tau_{21}}\right) \tag{4.19}$$

where $\tau_{21} \equiv (A_{21})^{-1}$c an be considered as the *decay time* or *lifetime* of level 2, as shown in Fig. 4.4. This decay time characterizes the statistical nature of spontaneous emission. It is also called the *fluorescence lifetime*.

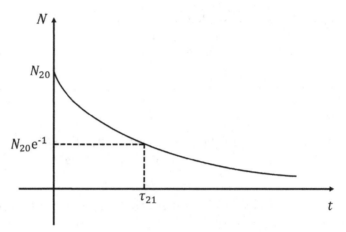

Fig. 4.4.   Spontaneous decay of level 2.

(4) Stimulated emission is a coherent process, i.e., the emitted radiation is in phase with the radiation that stimulates the emission.

(5) If we were to ask what the total rate of change of N_1 is without imposing the condition of thermal equilibrium, we would have, from Eqs. (4.9) to (4.12):

$$\frac{\partial N_1}{\partial t} = -B_{12}\rho(\nu)N_1 + B_{21}\rho(\nu)N_2 + A_{21}N_2. \qquad (4.20)$$

Using Eqs. (4.17) and (4.18), this becomes:

$$\frac{\partial N_1}{\partial t} = -\frac{g_2}{g_1}B_{21}\rho(\nu)N_1 + B_{21}\rho(\nu)N_2 + A_{21}N_2. \qquad (4.21)$$

Normally, spontaneous emission is rather weak and can be neglected as compared to stimulated emission. Equation (4.21) thus becomes, by setting $A_{21}N_2 \approx 0$:

$$\frac{\partial N_1}{\partial t} = B_{21}\rho(\nu)\left(N_2 - \frac{g_2}{g_1}N_1\right). \qquad (4.22)$$

This equation says that the net rate of change of the number of atoms in level 1 is equal to the probability per second of down transition ($B_{21}\rho(\nu)$) multiplied by the net number of atoms, leaving level 2 which is $N_2 - \frac{g_2}{g_1}N_1$.

4.3 Level Broadening

In reality, the physical world is more complicated. The atoms will interact with their environment. Even if they are isolated as in the previous assumptions, they still interact with the photons. Any of these interactions will lead to a width in a transition, i.e., broadening. For example, instead of having N_2 atoms in level 2 all having the same energy E_2, we have now, because of the broadening, the situation is shown in Fig. 4.5, where $g(\nu, \nu_0)$ is the *distribution function* or *lineshape function*. The net modification in the transition is shown in Fig. 4.6. The width of level 2 represents the *relative* change with respect to level 1 during the transition, regardless of absorption or emission. That is why level 1 is not broadened. Now,

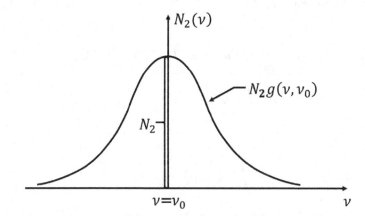

Fig. 4.5. Broadening of level 2.

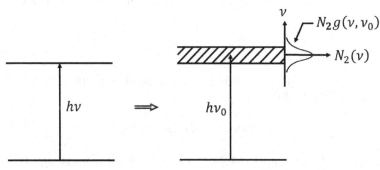

Fig. 4.6. Level broadening.

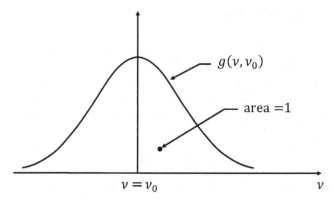

Fig. 4.7.   The distribution function $g(\nu, \nu_0)$.

$N(\nu) = N_2 g(\nu, \nu_0)$ by definition (Fig. 4.5). Integrating with respect to frequency, we have:

$$N_2 \equiv \int_0^\infty N(\nu)\mathrm{d}\nu = \int_0^\infty N_2 g(\nu, \nu_0)\mathrm{d}\nu,$$

$$= N_2 \int_0^\infty g(\nu, \nu_0)\mathrm{d}\nu.$$

Hence,

$$\int_0^\infty g(\nu, \nu_0)\mathrm{d}\nu = 1, \tag{4.23}$$

i.e., the area under $g(\nu, \nu_0)$ equals 1 as shown in Fig. 4.7. Physically, this can be understood as follows. The number of atoms in level 1 capable of absorbing photons in the range $h\nu$ to $h(\nu + \mathrm{d}\nu)$ is (see Fig. 4.5):

$$N_1(\nu)\mathrm{d}\nu = N_1 g(\nu, \nu_0)\mathrm{d}\nu. \tag{4.24}$$

Similarly, the number of atoms in level 2 capable of emitting photons in the range $h\nu$ to $h(\nu + \mathrm{d}\nu)$ is:

$$N_2(\nu)\mathrm{d}\nu = N_2 g(\nu, \nu_0)\mathrm{d}\nu. \tag{4.25}$$

From Eqs. (4.24) and (4.25), we see that (omitting ν_0 for compactness)

$g(\nu)$ = probability of absorption or emission per unit frequency

and

$g(\nu)\mathrm{d}\nu$ = probability of absorption or emission of photons in
the range $h\nu$ and $h(\nu + \mathrm{d}\nu)$

$\int_0^\infty g(\nu)\mathrm{d}\nu$ = total probability of transitions (absorption
and emission) between $\nu = 0$ and $\nu = \infty$.

And of course such a total probability in our isolated system must
be equal to 1, which is Eq. (4.23).

Physically, there are two classes of broadening: homogeneous and
inhomogeneous.

Definition: *Homogeneous broadening*

All atoms, during a transition, are affected in an identical way and
therefore the transitions of all the atoms are broadened identically.

Definition: *Inhomogeneous broadening*

Each atom's transition frequency is shifted to a different and distinct
extent so that the total broadening of the transitions is a "combination" of all these individual shifts.

Examples of homogeneous broadening:

(a) Natural linewidth:
From quantum mechanics, we know that a photon possesses
momentum. When an atom radiates a photon through spontaneous emission there is a back reaction exerted on the atom
by the photon. This leads to an atomic recoil, thus creating an
uncertainty in the position of the electron in the atom. This is
equivalent to saying that there is an uncertainty in the energy of
level 2. Using the Heisenberg uncertainty principle, we have:

$$(\Delta E)(\Delta t) \sim h.$$

Now, in order to make a measurement of the transition by spontaneous emission, we need to measure the emitted photon. Since

the lifetime of level 2 is τ_{21}, it takes about τ_{21} seconds to make a measurement,

$$\text{i.e.,} \quad \Delta t = \tau_{21},$$

$$\Delta E \sim \frac{h}{\tau_{21}},$$

i.e., the energy of the emitted photon can only be measured to an accuracy of $\Delta E \sim h/\tau_{21}$, or to within a width of

$$\Delta v = \frac{\Delta E}{h} \sim \frac{1}{\tau_{21}}.$$

(b) Collision broadening: If there is no collision, an ensemble of excited atoms will emit a long train of electromagnetic wave (Fig. 4.8(a)). Whenever there is a collision, the emission process is momentarily terminated so that the emitted wave train is shortened, i.e., the long wave train is truncated (Fig. 4.8(b)). From the wave theory, any truncated wave will have spectral sidebands. This means that there is a spectral width. Such collision broadening takes place implicitly at a constant temperature.

(c) Thermal broadening: Because temperature is really a measure of the average kinetic energy of all the particles in a system, changing temperature means changing the particle's kinetic energy. Its effect on transition broadening is essentially due to collisions; the higher the temperature is, the more collisions there are and the broader the transition is.

Examples of inhomogeneous broadening:

(a) Doppler broadening:
This is due to the atomic motion at a velocity v. Thus, each atomic transition is Doppler shifted from ν:

$$\nu_0 \to \nu_0 \left(1 + \frac{v}{c}\right), \quad \text{(if } v \ll c\text{).}$$

Because the ensemble of atoms normally has a distribution in the velocity space, v is different for different atoms so that the individual shifts of the transitions of all the atoms are different. The broadening is thus a "combination" of all these shifts.

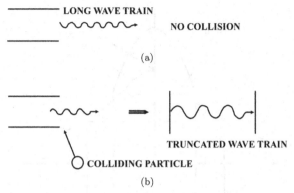

Fig. 4.8.   Collision broadening.

(b) Broadening due to crystal inhomogeneity:

In crystalline lasing materials (the newly invented verb "lase" is now popularly used to mean the action of emitting laser radiation), the active atoms are essentially ions doped uniformly into the lattice of a host crystal. For example, Cr^{+++} ions in a ruby crystal, which is aluminum oxide or Nd^{+++} ions in glass or YAG (Yttrium aluminum garnet). These active ions are not distributed perfectly uniformly in the host. Microscopically, an ion might "see" a different local environment due to crystal defects, random variations of dislocations, lattice strain, etc. Thus, each ion will experience a different local static electric field, which will shift the transition by a different extent. (We might call this a local Stark shift.) The combined result is the combination of all the different shifts, resulting in an inhomogeneous broadening.

Mathematical form of $g(\nu, \nu_0)$:

It is known that homogeneous broadening has a Lorentian distribution while inhomogeneous broadening has a Gaussian distribution. i.e.,

$$g(\nu, \nu_0) = \begin{cases} \dfrac{\Delta\nu}{2\pi}\left((\nu - \nu_0)^2 + \left(\dfrac{\Delta\nu}{2}\right)^2\right)^{-1} & \text{(Lorentian)} \\[4ex] \dfrac{2}{\Delta\nu}\left(\dfrac{\ln 2}{\pi}\right)^{1/2}\exp\left(-\left(\dfrac{\nu - \nu_0}{\Delta\nu/2}\right)^2 \ln 2\right) & \text{(Gaussian)} \end{cases}$$

(4.26)

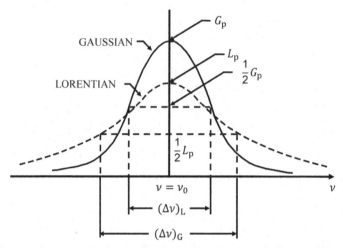

Fig. 4.9.   Schematic drawing comparing Lorentian and Gaussian distributions.

This is shown graphically in Fig. 4.9 where the peak values of the two distributions are.

$$L_p = g(\nu_0)_L = \frac{2}{\pi \Delta \nu} \quad \text{(Lorentian)} \tag{4.27}$$

$$G_p = g(\nu_0)_G = \frac{2}{(\Delta \nu)\pi}(\pi \ln 2)^{1/2} \quad \text{(Gaussian)} \tag{4.28}$$

4.4   Consequence of Broadening

The width $\Delta \nu$ of a transition is much broader than the width $d\nu$ of the laser line. (See Section 4.8 for more discussion.) For example, in the case of ruby laser, $\Delta \nu \sim 0.5$ nm and $d\nu \sim 0.01$ to 0.001 nm. The aforementioned isolated 2-level system in thermal equilibrium with the radiation field will then operate in a slightly different way. This is shown in Fig. 4.10. The laser is forced to operate at the frequency ν_l of width $d\nu$, while the lineshape (distribution) function of the transition is given by $g(\nu, \nu_0)$ of width $\Delta \nu \cdot (d\nu \ll \Delta \nu)$. Thus, the net number of atoms interacting with the laser radiation at ν_l of width $d\nu$ = (net number of atoms ready to leave level 2) × (probability of stimulated emission from ν_l to $\nu_l + d\nu$) = $(N_2 - \frac{g_2}{g_1}N_1) \cdot g(\nu_l, \nu_0)d\nu$ (see explanation of Eq. (4.22) and the meaning of $g(\nu, \nu_0)$).

Replacing the term $(N_2 - \frac{g_2}{g_1}N_1)$ in Eq. (4.22) by the above expression, we obtain the equation describing the rate of change of N_1 when

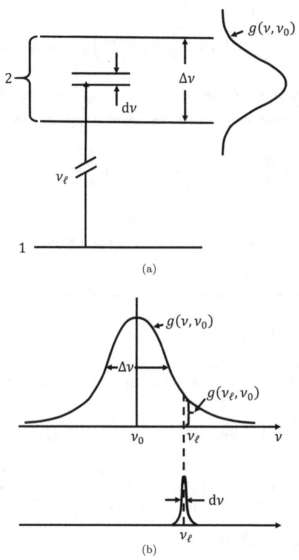

(a)

(b)

Fig. 4.10. Lasing transition at frequency ν_l of width $d\nu$ in a broad line shape $g(\nu, \nu_O)$ of width $\Delta\nu$.

interacting with the laser field:

$$\frac{\partial N_1}{\partial t} = B_{21}\rho(\nu_l)\left(N_2 - \frac{g_2}{g_1}N_1\right)g(\nu_l, \nu_0)d\nu.$$

Definition: <u>Number density</u> $n_i \equiv \frac{N_i}{V}$, $i = 1, 2$, and $V \equiv$ volume.

With this definition, dividing the above expression by $(-V)$, we obtain:

$$-\frac{\partial n_1}{\partial t} = B_{21}\rho(\nu_l)d\nu g(\nu_l, \nu_0)\left(\frac{g_2}{g_1}n_1 - n_2\right). \qquad (4.29)$$

But

$$-\frac{\partial n_1}{\partial t} \equiv \text{net rate of decrease of the density of atoms in level 1}$$

$$= \text{net absorption}$$

$$= \text{net rate of decrease of the laser radiation's photon density}$$

$$= -\frac{\partial}{\partial t}\left(\frac{\rho(\nu_l)d\nu}{h\nu_l}\right),$$

where it is assumed that all the laser photons have the same frequency ν_l (because $d\nu \ll \Delta\nu$). Substituting into Eq. (4.29), we get:

$$-\frac{\partial \rho(\nu_l)}{\partial t} = \rho(\nu_l)h\nu_l B_{21}g(\nu_l, \nu_0)\left(\frac{g_2}{g_1}n_1 - n_2\right). \qquad (4.30)$$

Consider that the laser photons pass through a slab of active material of width dx, as shown in Fig. 4.11. If c denotes the speed of the radiation in the material, the passage time is $dt = \frac{dx}{c}$.

$$\therefore -\frac{\partial \rho(\nu_l)}{\partial t} \rightarrow -\frac{\partial \rho(\nu_l)}{\partial x/c},$$

$$\therefore \text{Eq. 4.30}) \rightarrow -\frac{\partial \rho(\nu_l)}{\partial x} = \rho(\nu_l)h\nu_l B_{21}g(\nu_l, \nu_0)\left(\frac{g_2}{g_1}n_1 - n_2\right)\frac{1}{c}.$$

$$\text{Integrating} \rightarrow \rho(\nu_l) = \rho_0(\nu_l)\exp\left\{-h\nu_l g(\nu_l, \nu_0)B_{21}\left(\frac{g_2}{g_1}n_1 - n_2\right)\frac{x}{c}\right\}.$$

$$(4.31)$$

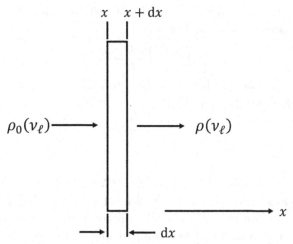

Fig. 4.11. Single pass amplification or attenuation through a general thin slab of optical medium.

Definition: Absorption coefficient

$$\alpha(\nu_l) \equiv \left(\frac{g_2}{g_1} n_1 - n_2 \right) \sigma_{21}(\nu_l), \qquad (4.32)$$

where

$$\sigma_{21}(\nu_l) \equiv \text{stimulated emission cross-section}$$

$$\equiv \frac{h\nu_l g(\nu_l, \nu_0)}{c} B_{21}. \qquad (4.33)$$

Substituting into Eq. (4.31) and by using Eq. (4.32), we have:

$$\rho(\nu_l) = \rho_0(\nu_l) \exp[-\alpha(\nu_l)x]. \qquad (4.34)$$

This is an expected result by inspection of Fig. 4.11. It is Beer's law if $\alpha(\nu_l) > 0$. However, if $\alpha(\nu_l) < 0$, the result will be:

$$\rho(\nu_l) = \rho_0(\nu_l) \exp[+|\alpha(\nu_l)|x],$$

which means amplification of the laser radiation at ν_l.

(*Note*: Similar to the definition of Eq. (4.33), we can also define the *absorption cross-section* as below.)

$$\sigma_{12}(\nu_l) = \frac{h\nu_l g(\nu_l, \nu_0)}{c} B_{12}. \tag{4.35}$$

Using the relationship between B_{12} and B_{21}, i.e., $B_{21} = \frac{g_1}{g_2}B_{12}$, comparing Eqs. (4.33) and (4.35):

$$\frac{\sigma_{12}}{\sigma_{21}} = \frac{g_2}{g_1}.$$

If

$$g_2 = g_1 = 1, \quad \Rightarrow \quad \sigma_{12} = \sigma_{21}.$$

This means that in a strictly 2-level system without degeneracy, the absorption and stimulated emission cross-sections are equal. (This relationship can also be obtained more rigorously using quantum mechanics.)

4.5 Impossibility of Having Gain (i.e., $\alpha(\nu_l) < 0$) in a 2-level System in Thermal Equilibrium

This can be seen by noting that in thermal equilibrium, from Eq. (4.14):

$$\frac{N_2}{N_1} = \frac{g_2}{g_1}\exp\left\{-\frac{(E_2 - E_1)}{kT}\right\} < \frac{g_2}{g_1}. \tag{4.36}$$

Hence,

$$N_2 < N_1\left(\frac{g_2}{g_1}\right). \tag{4.36'}$$

Or

$$N_1\left(\frac{g_2}{g_1}\right) - N_2 > 0.$$

Or

$$n_1\left(\frac{g_2}{g_1}\right) - n_2 > 0.$$

Thus,

$$\alpha(\nu_l) > 0, \quad \text{(using Eq. (4.32))}$$

and Eq. (4.34) is always an exponentially decreasing function. That is, there can be no net amplification of radiation in a 2-level system in thermal equilibrium.

Even if one increases the temperature of the system so as to increase N_2/N_1 (see Eq. (4.36)), the best one can have is $N_2 = N_1\left(\frac{g_1}{g_2}\right)$ at $T = \infty$. This leads to (by Eqs. (4.36) and (4.32)) $\alpha(\nu_\ell) = $ and the result is saturation, i.e., no gain and no loss ($\rho(\nu_\ell) = \rho_0(\nu_\ell)$ from Eq. (4.34)).

It is customary to use the temperature as a parameter to characterize the active medium. Thus, if by some means, one can make $\alpha(\nu_\ell) < 0$, one will have $\left(N_1\frac{g_2}{g_1} - N_2\right) < 0$ in Eq. (4.36′).

This is equivalent to making $T < 0$ in Eq. (4.36). One thus says that the active medium has achieved a *negative temperature*. (We should keep in mind that, physically, a negative temperature is impossible to achieve.)

Under such a situation of negative temperature, $N_2 > \frac{g_2}{g_1} N_1$, we say that the *population is inverted*, or we have an *inverted system*. In the case of $g_2 = g_1 = 1$, $N_2 > N_1$, *population inversion* means that there are more atoms in level 2 than level 1 in a 2-level system without degeneracy. *In order to achieve population inversion, one has to pump the system by some means other than heating.*

4.6 Pumping

One can imagine using some "magic" means to raise atoms from level 1 to level 2 so as to reach population inversion. However, if the system is to be in equilibrium, isolated and diluted (no collision), the only means to do so is by absorption of radiation. We thus fall back to our initial conditions set forth at the beginning of this chapter, i.e., thermal equilibrium with the radiation at the transition frequency. And we already know that inversion is impossible.

In practice, one pumps atoms into level 2 via some other levels. One can either raise atoms from level 1 into level 2 via a third level (3-level system) or raise atoms from a third, irrelevant level into level 2 via a fourth also irrelevant level (4-level system). Most known laser systems operate with a 4-level system while 3-level systems are used to a much lesser extent because of the inefficiency to attain inversion (see below).

4.7 Rate Equations Approach

We now go into some detail of 3- and 4-level systems using the intuitive rate equation approach. The rate equation approach essentially balances the rate of gain and loss of particles (atoms) in different levels and also of the laser photons. Monochromaticity of the laser is assumed while longitudinal and spatial distributions of the laser radiation inside the laser cavity are ignored. This means that we are assuming that the photon density is uniform inside the cavity so that we can analyze only the change along the axis of the laser. Such an approximation is reasonable because the spatial (longitudinal and transverse) distribution of the laser radiation (or modes) inside the cavity depends mostly on the geometry of the cavity and, in particular, the size, shape and aperture of the end mirrors. The active medium is just an amplifier that enhances the radiation densities inside those electromagnetic modes satisfying the boundary conditions inside the cavity. Such a decoupling of the active medium and the cavity makes life easier in the analysis of laser oscillation. One can study first of all the energy and particle number balances (rate equation analysis) and the mode structures independently. After that, the results can be matched, giving the realistic results.

We now concentrate on the rate equations. The result should give us broad features of the average power, peak power, laser temporal pulse envelope, and threshold conditions, etc.

(a) Idealized three level system:

The system is shown in Fig. 4.12. One pumps atoms in level 1 into level 3, which is rather broad. This pumping could be the absorption of some appropriate radiation from a flashlamp, another laser, etc. It could also be achieved by forward bias in the p–n junction, collision, chemical reaction, etc. Whatever it is, let us ignore the detail and assume simply that atoms have been raised from level 1 into level 3 at a rate W_p. On reaching level 3, they decay rapidly onto level 2 and stay there for a long enough time to allow laser action to occur between levels 2 and 1. We still keep the previous assumptions that the system is isolated and in equilibrium and that no collision among the atoms is allowed. Thus, the decay mechanism of levels 3 and 2 is through spontaneous emission whose lifetimes are τ_{31}, τ_{32} and τ_{21}. The above statement of the rapid decay from 3→2 means the following additional assumptions:

$$\tau_{31} \gg \tau_{32}, \tag{4.37}$$

$$\tau_{21} \gg \tau_{32}, \tag{4.38}$$

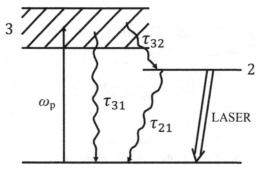

Fig. 4.12. A 3-level system.

$$n_3 \approx 0. \tag{4.39}$$

Hence,

if $n_{\text{tot}} \equiv$ total number of atoms in the system.

We have

$$n_1 + n_2 + n_3 = n_{\text{tot}},$$

$$n_1 + n_2 \approx n_{\text{tot}}, \tag{4.40}$$

Under these assumptions, we derive the following rate equations:

$$\frac{\partial n}{\partial t} = -\gamma\sigma_{21}c\phi n - \frac{(\gamma - 1)n_{\text{t0t}} + n}{\tau_f} + W_p(n_{\text{tot}} - n) \tag{4.41}$$

and

$$\frac{\partial \phi}{\partial t} = \sigma_{21}c\phi n - \frac{\phi}{\tau_c} + S, \tag{4.42}$$

where

$$n \equiv \text{inversion density},$$

$$\equiv n_2 - \frac{g_2}{g_1}n_1, \tag{4.43}$$

$$\gamma \equiv 1 + \frac{g_2}{g_1}. \tag{4.44}$$

$\tau_f \equiv \tau_{21}$, and $\tau_c \equiv$ *photon lifetime* in the cavity (see below) and $\phi =$ photon density (cm^{-3}). S is the spontaneous emission rate.

Derivation:

From the definition of the stimulated emission cross-section σ_{21} (Eq. (4.33)), we have:

$$B_{21} = \frac{c}{h\nu g(\nu)}\sigma_{21}(\nu). \qquad (4.45)$$

Now,

$$\rho(\nu) \equiv \text{laser radiation density/frequency,}$$

$$\equiv h\nu\phi \cdot (\text{net probability of stimulated emission/}$$

$$\text{frequency),}$$

$$\equiv h\nu\phi \cdot g(\nu).$$

Hence,

$$B_{21}\rho(\nu) = c\sigma_{21}(\nu)\phi. \qquad (4.46)$$

Now,

$$\frac{\partial n_1}{\partial t} = \text{net stimulated emission rate}$$

$$+ \text{spontaneous emission rate}$$

$$- \text{pumping rate of level 1 into level 3}$$

$$= \left(n_2 - \frac{g_2}{g_1}n_1\right)B_{21}\rho(\nu) + \frac{n_2}{\tau_{21}} - W_p n_1, \qquad (4.47)$$

where the first term comes from Eq. (4.22).
Now:

$$\frac{\partial n}{\partial t} \equiv \frac{\partial}{\partial t}\left(n_2 - \frac{g_2}{g_1}n_1\right),$$

$$= -\frac{\partial n_1}{\partial t}\left(1 + \frac{g_2}{g_1}\right), \qquad (4.48)$$

$$\left(\text{since } n_1 + n_2 \approx n_{\text{tot}} = \text{const.; } \frac{\partial n_1}{\partial t} = -\frac{\partial n_2}{\partial t}\right).$$

Substituting Eqs. (4.46) and (4.47) into Eq. (4.48), and using Eqs. (4.43) and (4.44), we obtain Eq. (4.41).

To obtain the photon rate equation:

$$\frac{\partial \phi}{\partial t} = \text{rate of creation of photon density} - \text{rate of loss},$$

$$= \text{net rate of stimulated emission} + \text{rate of spontaneous emission}$$

$$- \text{rate of total loss in the cavity},$$

$$= \left(n_2 - \frac{g_2}{g_1} n_1 \right) B_{21} \rho(\nu) + S - \frac{\phi}{\tau_c}, \tag{4.49}$$

where S is the spontaneous emission rate and τ_c is a decay constant of the photon density. τ_c can be related to Q and α_c of the cavity by the following consideration. Assuming there is no emission and absorption, and a photon density ϕ_0 is injected by some means into the laser cavity. ϕ will decay through all the possible loss mechanisms mentioned at the beginning of the chapter.

Hence,

$$\frac{\partial \phi}{\partial t} = -\frac{\phi}{\tau_c},$$

$$\phi = \phi_0 e^{-t/\tau_c}. \tag{4.50}$$

But

$$E = E_0 e^{-\frac{\omega}{Q} t} \quad (\text{Eq. } (4.4)),$$

the two equations are identical because the photon distribution is assumed uniform in the cavity. Hence, using also Eq. (4.4'):

$$\tau_c = \frac{Q}{\omega} = \frac{1}{\alpha_c c}. \tag{4.51}$$

τ_c is called the *photon lifetime* in the cavity. Using Eq. (4.22), we can rewrite Eq. (4.49):

$$\frac{\partial \phi}{\partial t} = \sigma_{21} c \phi n - \frac{\phi}{\tau_c} + S,$$

which is Eq. (4.42).

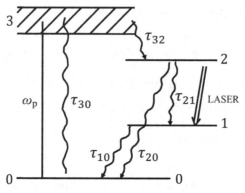

Fig. 4.13.   A 4-level system.

(b) Idealized 4-level system:

This is shown in Fig. 4.13. Again, we ignore for the moment the detail of pumping. In addition to the assumption of equilibrium, no collision and isolation, we further assume:

$$\tau_{30} \gg \tau_{32},$$

$$\tau_{21} \gg \tau_{32},$$

$$\tau_{20} \gg \tau_{32},$$

$$\tau_{10} \approx 0,$$

i.e., after a fast pumping of atoms from level 0 to level 3, the atoms decay very rapidly into level 2 and wait there till laser action takes place between $2 \to 1$. Once arriving at level 1, the atoms "immediately" decay back to level zero. Hence,

$$n_3 \approx 0,$$

$$n_1 \approx 0 \qquad\qquad (4.52')$$

and

$$n_0 + n_1 + n_2 + n_3 = n_{\text{tot}} \text{ becomes}$$

$$n_0 + n_2 = n_{\text{tot}} = \text{constant}, \qquad\qquad (4.52)$$

where $n_{\text{tot}} \equiv$ total number density of atoms in the system.

Definition: $\frac{1}{\tau_f} \equiv \frac{1}{\tau_{21}} + \frac{1}{\tau_{20}}$.

Again, we derive the following rate equations:

$$\frac{\partial n}{\partial t} = W_p(n_{\text{tot}} - n) - n\sigma_{21}\phi c - \frac{n}{\tau_f} \qquad (4.53)$$

and

$$\frac{\partial \phi}{\partial t} = \sigma_{21}c\phi n - \frac{\phi}{\tau_c} + S. \qquad (4.54)$$

Derivation:

$\frac{\partial n_2}{\partial t}$ = pumping – net stimulated emission – spontaneous emission, =
$W_p n_0 - \left(n_2 - \frac{g_2}{g_1}n_1\right)\sigma_{21}\phi c - \frac{n_2}{\tau_{21}} - \frac{n_2}{\tau_{20}}$.

   Since

$$n \equiv n_2 - \frac{g_2}{g_1}n_1 \approx n_2, \quad (\text{since } n_1 \approx 0), \qquad (4.55)$$

$$\therefore \frac{\partial n}{\partial t} = W_p n_0 - n\sigma_{21}\phi c - \frac{n_2}{\tau_f},$$

which is Eq. (4.54) after using Eqs. (4.52) and (4.55).

Equation (4.54) can be derived in an identical way as that leading to Eq. (4.42) and is left as an <u>exercise</u> to the reader.

4.8   Threshold Oscillation

Now Eqs. (4.34) and (4.32) show that the gain of the active medium should be given by:

$$g = -\alpha(\nu_\ell) = \left(n_2 - \frac{g_2}{g_1}n_1\right)\sigma_{21}(\nu_\ell). \qquad (4.55')$$

But the threshold oscillation condition is (Eq. (4.3)):

$$gl_m = \alpha_c l.$$

Hence,

$$-\alpha(\nu_l)l_m = \alpha_c l,$$

$$\left(n_2 - \frac{g_2}{g_1}n_1\right)\sigma_{21}(\nu_\ell)l_m = \alpha_c l.$$

After some rearrangement, we obtain:

$$n \equiv n_2 - \frac{g_2}{g_1} n_1 = \frac{8\pi\nu^2 \tau_{21} l}{\tau_c g(\nu_\ell, \nu_0) c^3 l_m}, \qquad (4.56)$$

where Eqs. (4.51), (4.33) and (4.17) have been used. Equation (4.56) shows that at *threshold* oscillation, the inversion n is inversely proportional to $g(\nu, \nu_0)$. If the laser is allowed to oscillate at any frequency ν_ℓ within the transition line width of $g(\nu, \nu_0)$, then once the active medium is pumped, all these frequencies will compete with one another to extract energy from the gain medium. The frequency at $\nu = \nu_0$ will first reach the threshold condition because the probability of net stimulated emission, which is $g(\nu, \nu_0)$ is largest at $\nu = \nu_0$, i.e., at the peak of $g(\nu, \nu_0)$. If the system is maintained at a threshold, once it starts oscillating at $\nu = \nu_0$, the other frequencies will be suppressed to oscillate because the energy in the gain medium is extracted by $\nu = \nu_0$ and there is not enough left for the other frequencies to overcome the loss. The laser thus oscillates at $\nu = \nu_0$ with a very narrow width. *That is why laser lines are very narrow.* Figure (4.14) shows schematically that at the peak of the gain curve, gain just overcomes loss, and laser action takes place only within the very narrow width $d\nu$.

One can, of course, force the laser to oscillate at any frequency $\nu \neq \nu_0$ within the width of $g(\nu, \nu_0)$ by introducing more loss to all frequencies except the one of interest. Then, net gain will build up only for the frequency of interest and it will oscillate first, and be maintained at threshold so that the other frequencies cannot oscillate.

It is, of course, desirable to have an efficient laser, i.e., as low an inversion n as possible. From Eq. (4.56), n is inversely proportional to τ_c, the photon density lifetime. That is, the longer τ_c is, the lower n will be. Long τ_c means small α_c according to Eq. (4.51), i.e., low loss.

We now ask the question: "What should the pumping rate ω_p be in order to maintain the laser at threshold oscillation?" For a 3-level system, the answer is:

$$W_p(\text{min.}) = \frac{g_2}{\tau_f g_1} \qquad (4.57)$$

and for a 4-level system, there is no minimum W_p, i.e., any W_p is good.

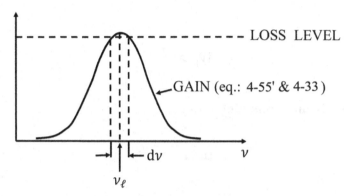

Fig. 4.14.   Threshold oscillation leading to a very sharp lasing line.

Proof:

(A) 3-level system
Assumptions:
(1) low threshold value:

$$\phi \approx 0.$$

(2) steady state condition:

$$\frac{\partial n}{\partial t} = 0.$$

Thus, Eq. (4.41) becomes:

$$0 = -0 - \frac{(\gamma - 1)n_{\text{tot}} + n}{\tau_f} + W_p(n_{\text{tot}} - n).$$

Rearranging terms, using Eq. (4.44), we obtain:

$$\frac{n}{n_{\text{tot}}} = \frac{W_p\tau_f - g_2/g_1}{W_p\tau_f + 1}. \qquad (4.58)$$

In order to have an inversion, we should have:

$$\frac{n}{n_{\text{tot}}} \geq 0.$$

From Eq. (4.58), this means:

$$W_p\tau_f - g_2/g_1 \geq 0$$

or

$$W_{\mathrm{p}} \geq \frac{g_2}{\tau_{\mathrm{f}} g_1}$$

and the minimum pumping rate is:

$$W_{\mathrm{p}}(\mathrm{min.}) \geq \frac{g_2}{\tau_{\mathrm{f}} g_1}, \tag{4.59}$$

which is Eq. (4.57). We can now take note of some of the disadvantages of a 3-level system.

Disadvantage 1:

From Eq. (4.59), we see that to obtain a low pumping rate W_p, one has to have a long τ_{f}, i.e., a long fluorescence lifetime for level 2. This would mean that other loss mechanisms might intervene, such as collisions that would de-excite level 2, resulting in lower efficiency.

Disadvantage 2:

At minimum inversion:

$$n = 0,$$

$$\text{i.e.,} \quad n_2 - \frac{g_2}{g_1} n_1 = 0$$

or

$$n_2 = \frac{g_2}{g_1} n_1 \tag{4.60}$$

If

$$g_2 = g_1 = 1, \quad n_2 = n_1 = \frac{1}{2} n_{\mathrm{tot}}, \tag{4.61}$$

this means that for non-degenerate levels 1 and 2, one needs to pump half of the total atoms into level 2 in order to just reach the minimum threshold and this means a lot of pumping, i.e., not efficient.

(B) 4-level systems

The same assumptions of

$$\phi \approx 0,$$

$$\frac{\partial n}{\partial t} = 0$$

are used. Thus, Eq. (4.53) becomes:

$$0 = W_p(n_{\text{tot}} - n) - 0 - \frac{n}{\tau_f},$$

$$\frac{n}{n_{\text{tot}}} = \frac{W_p}{W_p} + \frac{1}{\tau_f}, \tag{4.62}$$

and the right hand side of Eq. (4.62) is always positive. Hence,

$$\frac{n}{n_{\text{tot}}} > 0 \quad \text{(always)}, \tag{4.63}$$

i.e., there is no minimum pump rate. Any W_p will induce an inversion in an idealized 4-level system. Physically, this is evident because n_1 was assumed zero (Eq. (4.52$'$)) so that whenever there is a slight population n_2 in level 2, it is inverted with respect to level 1.

4.9 Threshold Pump Power

Essentially, at threshold oscillation, the pumping is to maintain $\phi \approx 0$ for the laser output, i.e., the pumping is mainly used to compensate for the loss in the cavity.

Assumption: The only loss in the cavity is through fluorescence or spontaneous emission.

If $P_{\text{eff}} \equiv$ effective pump power,

$P_f \equiv$ fluorescence power.

Our assumption means:

$$P_{\text{eff}} = P_f,$$

$$= \frac{n_2(th) \cdot h\nu}{\tau_f},$$

where $n_2(th)$ threshold population of level 2. For idealized 3- and 4-level systems, we have:

$$
P_{eff} =
\begin{cases}
\dfrac{h\nu}{\tau_f} \cdot \dfrac{n_{\text{tot}}}{2} & \text{(3-level, using Eq. (4.61))} \qquad (4.64) \\[2ex]
\dfrac{h\nu}{\tau_f} n & \text{(4-level, using Eq. (4.55))} \qquad (4.65)
\end{cases}
$$

4.10   Above Threshold Oscillation and Gain Saturation

Very often, a laser oscillates with a photon flux above that of threshold, i.e., $\phi \neq 0$ but $\frac{\partial n}{\partial t} = 0$ (still). This means that the steady-state condition is still valid but in the presence of a strong ϕ.

3-level system:

The rate Eq. (4.41) now becomes:

$$0 = -\gamma \sigma_{21} c \phi n - \frac{(\gamma - 1)n_{\text{tot}} + n}{\tau_f} + W_{\text{p}}(n_{\text{tot}} - n).$$

Solving for n, we obtain:

$$n = n_{\text{tot}} \left[W_{\text{p}} - \frac{\gamma - 1}{\tau_f} \right] \left(\gamma \sigma_{21} c \phi + W_p + \frac{1}{\tau_f} \right)^{-1} \quad \text{(3-level)}.$$

$$(4.66)$$

4-level system:

The rate Eq. (4.53) becomes:

$$0 = W_p(n_{\text{tot}} - n) - n\sigma_{21}\phi c - \frac{n}{\tau_f}.$$

Solving for n, we obtain:

$$n = n_{\text{tot}} \cdot \frac{W_p}{W_p + \sigma_{21}\phi c + \frac{1}{\tau_f}} \quad \text{(4-level)}. \qquad (4.67)$$

We now define the small gain coefficient g_0. This follows from the expression of the gain coefficient g given by (using Eq. (4.32)):

$$g = -\alpha(\nu_l) = \left(n_2 - \frac{g_2}{g_1}n_1\right)\sigma_{21}(\nu_\ell t)$$

or

$$g = n\sigma_{21} \quad \text{(using Eq. (4.43))}. \qquad (4.68)$$

Definition: $g_0 = g \, (\text{at } \phi\neq)$,
$$= n(\phi \neq 0)\sigma_{21} \, (\text{from Eq. (4.68)}),$$

$$= \begin{cases} \sigma_{21}n_{\text{tot}}\left(W_p - \frac{\gamma-1}{\tau_f}\right)\left(W_p + \frac{1}{\tau_f}\right)^{-1} & \text{(3-level)} \qquad (4.69) \\[2em] \sigma_{21}n_{\text{tot}}W_p\left(W_p + \frac{1}{\tau_f}\right)^{-1} & \text{(4-level)}. \qquad (4.70) \end{cases}$$

From Eqs. (4.69) and (4.70), we see that g_0 depends only on the material parameters (σ_{21} and τ_f) and the pumping rate W_p. This small signal gain is what an active medium has when it is pumped above the threshold and when the laser action is inhibited (e.g., by blocking the light inside the cavity, increasing the loss significantly during a period of time, i.e., Q-switching, etc.). If feedback is restored at some moment, ϕ in the resonator will increase exponentially in the beginning, i.e., the increase follows:

$$e^{g_0 x}.$$

As soon as the ϕ becomes appreciable, g_0 becomes g where $g = n\sigma_{21}$, (from Eq. (4.68)),

$$
= \begin{cases}
\sigma_{21}n_{tot}\left(W_p - \dfrac{\gamma - 1}{\tau_f}\right)\left[\left(W_p + \dfrac{1}{\tau_f}\right)\left(\dfrac{\gamma\sigma_{21}c\phi}{W_p + \frac{1}{\tau_f}} + 1\right)\right]^{-1} \\
\hspace{6cm}\text{(3-level)} \qquad (4.71) \\[2em]
\sigma_{21}n_{tot}W_p\left[\left(W_p + \dfrac{1}{\tau_f}\right)\left(1 + \dfrac{\sigma_{21}\phi c}{W_p + \frac{1}{\tau_f}}\right)\right]^{-1} \quad \text{(4-level)} \\
\hspace{10cm} (4.72)
\end{cases}
$$

Using Eqs. (4.69) and (4.70), we simplify Eqs. (4.71) and (4.72) into:

$$
g = \begin{cases}
g_0\left(1 + \dfrac{\gamma\sigma_{21}\phi c}{W_p + \frac{1}{\tau_f}}\right)^{-1} & \text{(3-level)} \qquad (4.73) \\[2em]
g_0\left(1 + \dfrac{\sigma_{21}\phi c}{W_p + \frac{1}{\tau_f}}\right)^{-1} & \text{(4-level)} \qquad (4.74)
\end{cases}
$$

Definition: *Intensity* $I \equiv c\phi h\nu \left(\dfrac{\text{Joules}}{\text{cm}^2\text{sec}}\right).$ $\qquad$ (4.75)

Definition: *Saturation intensity* $\equiv I_s,$

$$\equiv \text{intensity at which } g = \frac{1}{2}g_0.$$

Since g is given by Eqs. (4.73) and (4.74), we have, at $I = I_s$:

$$
\frac{1}{2}g_0 = \begin{cases}
g_0\left(1 + \dfrac{\gamma\sigma_{12}\phi_s c}{W_p + \frac{1}{\tau_f}}\right)^{-1} & \text{(3-level)} \qquad (4.76) \\[2em]
g_0\left(1 + \dfrac{\sigma_{21}\phi_s c}{W_p + \frac{1}{\tau_f}}\right)^{-1} & \text{(4-level)} \qquad (4.77)
\end{cases}
$$

where $\phi_s = $ *saturation photon density at* $I = I_s$.

From Eq. (4.75), using Eqs. (4.76) and (4.77), one obtains

$I_s \equiv c\phi_s h\nu,$

$$= \begin{cases} \left(W_p + \dfrac{1}{\tau_f}\right) \dfrac{h\nu}{\gamma\sigma_{21}} & \text{(3-level)} \qquad (4.78) \\[3mm] \left(W_p + \dfrac{1}{\tau_f}\right) \dfrac{h\nu}{\sigma_{21}} & \text{(4-level)} \qquad (4.79) \end{cases}$$

Substituting Eqs. (4.78) and (4.79) into Eqs. (4.73) and (4.74), we have the "universal" relation:

$$g = \frac{g_0}{1 + I/I_s} \qquad \text{(both 3- and 4-level systems).} \qquad (4.80)$$

Definition: $r_{st} \equiv$ stimulated emission lifetime

The notion of stimulated emission lifetime is sometimes used. We explain it as follows. Assume that a population n_2 is generated in level 2 and left alone in a gain medium. Its evolution (decay) in time will include fluorescence and stimulated emission:

$$\frac{\partial n_2}{\partial t} = -\frac{n_2}{\tau_f} - \frac{n_2}{\tau_{st}}. \qquad (4.81)$$

But

$$\frac{\partial n_2}{\partial t} = -\frac{n_2}{\tau_f} - \text{stimulated emission},$$

$$= -\frac{n_2}{\tau_f} - B_{21}\rho(\nu)n_2. \qquad (4.82)$$

Equations (4.81) and (4.82) give

$$\frac{1}{\tau_{st}} = B_{21}\rho(\nu) = c\sigma_{21}\phi, \qquad \text{(from Eq. (4.46)),}$$

$$\tau_{st} = \frac{1}{\sigma_{21}c\phi} = \frac{h\nu}{\sigma_{21}I}, \qquad \text{(from Eq. (4.75)).} \qquad (4.83)$$

Physically, Eq. (4.83) shows that the stimulated emission lifetime of level 2 is inversely proportional to both the stimulated emission cross-section and intensity at a fixed frequency. The larger σ_{21} and I are, the shorter τ_{st} will be. Thus, in the presence of strong stimulated emission (large σ_{21} and I):

$$\tau_f \gg \tau_{st}$$

and we can neglect the first term (spontaneous emission) on the right hand side of Eq. (4.81). In other words, in most practical laser calculations, spontaneous emission is neglected. (Cf. Eq. (4.22) and the discussion therein.)

4.11 Output Power Calculation

We now calculate the output power from a laser oscillator, taking into account possible saturation in the gain medium. We start by considering the amplification of intensity in a single pass through the gain medium. An oscillation condition is then imposed to obtain the equation for the output power.

(a) Single pass amplification

We first ask what the amplification of a beam of light is after a single pass through the amplifying medium of the laser. Referring to Eqs. (4.34) and (4.55') and Fig. 4.11, we see that the gain in radiation density $\rho(\nu_l)$ across an amplifying "slab" of thickness dx (Fig. 4.15(a)) is

$$\rho(x + dx) - \rho(x) \equiv d\rho(x),$$
$$= (g\,dx)\rho(x), \tag{4.84}$$

where g, the gain, is given by Eq. (4.55'). It is understood that the laser operates at the frequency ν_ℓ so that we omit writing it in the argument of $\rho(x)$. From Eq. (4.75), the intensity

$$I \equiv c\phi h\nu = c\rho. \tag{4.85}$$

Thus, multiplying Eq. (4.84) by c and using Eq. (4.85), we have

$$dI(x) = (g\,dx)I(x). \tag{4.86}$$

If we include the loss in Eq. (4.86), we have

$$dI(x) = (g - \alpha_c)(dx)I(x), \tag{4.87}$$

where α_c is the loss coefficient (see Section 4.1). Equation (4.87) becomes

$$\frac{dI(x)}{dx} = (g - \alpha_c)dx. \tag{4.88}$$

Note that g is now given by Eq. (4.80), which we rewrite as follows:

$$g = \frac{g_0}{1 + I/I_S}.$$ (4.89)

Substituting Eq. (4.89) into (4.88), we have

$$\frac{dI(x)}{dx} = \frac{g_0 I(x)}{1 + I(x)/I_S} - \alpha_c I(x),$$ (4.90)

which is the *amplifier equation*. The analytical solution is not available. Numerical integration is normally required.

The condition for *small signal gain* can be defined as:

$$I(x) \ll I_S,$$ (4.91)

$$g \approx g_0.$$ (4.92)

Equation (4.90) becomes

$$\frac{dI(x)}{dx} = g_0 I(x) - \alpha_c I(x)$$

or

$$I(x) = I_0 e^{(g_0 - \alpha_C)x},$$ (4.93)

which is essentially Eq. (4.2). Here, I_0 is the intensity at $x = 0$, and x is the position at a point along the amplifier's axis (Fig. 4.15(b)). This shows also that the introduction in Section 4.1 pertains only to the case of small signal gain.

(b) Oscillation

We now allow oscillation to take place and simplify the laser cavity by assuming that the ends of the amplifying medium also act as reflectors, plane parallel in the present analysis. This is shown in Fig. 4.15(c). The mirror at $x = 0$ is assumed to be one that takes care of the total loss in the cavity (cf. Section 4.1), while the mirror at $x = \ell$ is the output mirror.

A beam of light "circulates" in the cavity as shown schematically in Fig. 4.15(c). $I_1(x)$ is the beam propagating towards the right side. When it reaches the mirror, part of it will be transmitted and partly reflected. All other losses are incorporated in the other mirror. Hence, the reflectivity of the mirror at $x = \ell$ is

$$R = \frac{I_2(\ell)}{I_1(\ell)}.$$ (4.94)

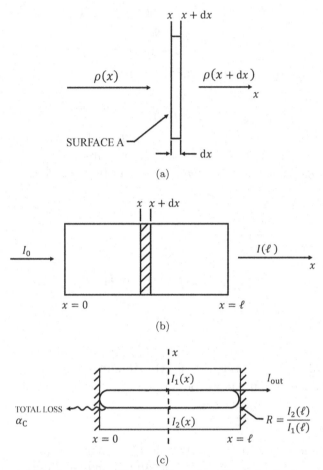

Fig. 4.15. (a) Single pass amplification through a thin slab of gain medium. (b) Single pass amplification through a general gain medium. (c) Above-threshold oscillation analysis.

When I_2 reaches the mirror at $x = 0$, part of it is lost into the mirror, and this loss represents the total loss α_C. The reflected beam should be equal to I_1 under the oscillation condition. We note that at any position x, the gain medium "sees" a total intensity I_T, which is the sum of I_1 and I_2.

$$\text{i.e.,} \quad I_T(x) = I_1(x) + I_2(x). \qquad (4.95)$$

The loss "seen" by each beam is α_C. The gain "seen" by each beam is

$$g_1 = g_2 = \frac{g_0}{1 + I_T/I_S}. \tag{4.96}$$

Here, g_1 and g_2 are the gains in both beams, not atomic degeneracies.

Note that it is the total intensity that saturates the gain medium at any point in the medium and thus we use I_T in the expression of the two gains. The coupled equations governing each beam for one pass are then (using Eq. (4.88)):

$$\frac{dI_1(x)}{dx} = g_1 I_1(x) - \alpha_C I_1(x), \tag{4.97}$$

$$-\frac{dI_2(x)}{dx} = g_2 I_2(x) - \alpha_C I_2(x). \tag{4.98}$$

The reason why we put a negative sign on the left hand side of Eq. (4.98) is that $I_2(x)$ propagates in the negative "x" direction. It can be seen that the *mean intensity* I in the cavity is constant, where

$$\underline{\text{Mean intensity }} I(x) \equiv \sqrt{I_1(x)I_2(x)}. \tag{4.99}$$

This can be proved by multiplying Eq. (4.97) by $I_2(x)$ and Eq. (4.98) by I_1 and taking the difference (exercise). The result is

$$\frac{dI(x)}{dx} = 0, \tag{4.100}$$

$$\text{so that} \quad I(x) = \text{constant}. \tag{4.101}$$

One can express the power output of the laser in terms of I, R and A. (A is the cross-sectional area of the gain medium.) This is left as an exercise. The result is

$$\text{Power } P \text{ (Joule/sec)} = AI \left(\frac{1 - R}{R^{1/2}} \right). \tag{4.102}$$

The precise values of I and P depend on the solution of the coupled Eqs. (4.97) and (4.98). A numerical solution is possible once g_0 and I_s are known. We will not go into any details here.

4.12   Different Types of Laser

We have so far discussed how to generate a laser in a resonant cavity
with a 3- or 4-level gain medium. There are three basic elements for
a laser: active medium, pumping source, and resonant cavity. For an
intuitive understanding of the principle of lasers, we will introduce
three typical lasers in this section. Depending on the gain medium
used, one distinguishes solid, liquid and gas state lasers. Furthermore, in the time domain, one distinguishes continuous wave (cw)
and pulsed lasers, and in tunability, one distinguishes fixed frequency
and tunable lasers. Other types of lasers will be discussed later in the
book.

Solid-state laser

In 1960, the first laser, the ruby laser, was invented by Theodore
Maiman. Research into solid lasers has continued for more than 50
years. Gradually significant progress has been reached in obtaining
super high power, ultrashort pulse duration, and tunable wavelengths
of laser radiation. Rudy is an Al_2O_3 crystal with a small amount of
Cr_2O_3 (about 0.05%), and the ruby laser is a solid-state 3-level laser;
the level and transition diagram is shown in Fig. 4.16, where only the
Cr^{3+} ions participate in the pumping and lasing processes. When the
ruby crystal is pumped by a flash lamp, many of the chromium ions
will be excited to the $^4F_1$ and $^4F_2$ states by absorbing the photons
corresponding to wavelengths of $\sim$400 and $\sim$550 nm, respectively.

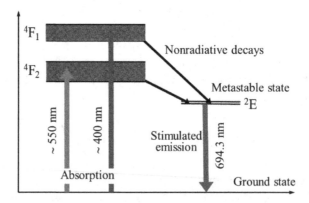

Fig. 4.16.   Level and transition diagram for the ruby laser.

Then the excited chromium ions decay non-radiatively to the ^{2}E state within 100 ns. Because of the very long lifetime (about 5 ms) of the metastable state ^{2}E, population inversion between the ^{2}E state and the ground state is quickly established, resulting in a fixed frequency 694.3-nm laser if the ruby crystal is located in a proper oscillator cavity.

It should be stressed that when stimulated emission takes place, the chromium ions in the ^{2}E state will fall to the ground state, leading to the disappearance of the population inversion. Thus, if the ruby crystal is continuously pumped, population inversion will be repeatedly built up, resulting in a sequence of pulsed light emissions, that is, a laser in a pulsed mode. When Q-switching is applied, such light spikes can be controlled, producing much stronger laser pulses in a designed repetition rate. Note that the 3-level ruby laser requires high energy pumping to deplete the population in the ground state. The success of the ruby laser is based on the broadband absorption of the ^{4}F$_1$ and ^{4}F$_2$ states, which efficiently utilize the pumping energy. As aforementioned, 4-level lasers, in which a final level is not the ground state, require much less power to build up and sustain the population inversion. Today, many solid-state 4-level lasers have been developed, which can work in both pulsed and cw output modes. One of such examples is the Nd: YAG (Nd^{3+}: Y$_3$Al$_5$O$_{12}$) laser.

Gas laser

A gas laser uses gas or vapor as a gain medium. As compared with a solid laser, a gas laser can output even higher average power. The laser is normally pumped by a collision scheme, which is widely used in industrial and agricultural production, defense and scientific research. The Helium–Neon (He–Ne) laser was the first gas laser built shortly after the ruby laser. It is a typical 4-level laser with the level and transition diagram shown in Fig. 4.17. Both the He and Ne atoms in the active gas mixture (typically with a ratio of 5:1~10:1) participate in the laser process. He atoms are excited to the states 2 ^{1}S and 2 ^{3}S by an electric discharge through electron impact. These two states, having almost the same energies as the 5S and 4S states of Ne, respectively, are metastable states with very long lifetimes. By collision with Ne atoms, the metastable He atoms transfer the energy to Ne atoms, so that Ne atoms are excited to the 5S and 4S states. Since the 4P and 3P states of Ne are normally unoccupied in the electric discharge, a population inversion

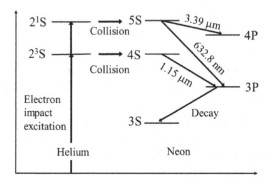

Fig. 4.17.   Level and transition diagram for the He–Ne laser.

between the S and the lower-lying P levels is naturally achieved, and thus stimulated emission at several different fixed wavelengths can be obtained, depending on the cavity selection.

The lifetimes of the 5S and 4S states ($\sim$100 ns) are much longer than those of the 4P and 3P states ($\sim$10 ns). When the gas mixture is continuously pumped, it is possible to maintain the population inversion and produce the He–Ne laser in a cw mode. Due to the low pressure of the He–Ne gas mixture (normally a few Torr), the He–Ne gain is normally small, resulting in low output powers, typically in a few mW. Today many gas-state 4-level lasers are available, such as the CO_2 laser that can work in both pulsed and cw output modes.

Dye lasers

Dye lasers are typical liquid-state 4-level laser systems and have been the first broadly tunable lasers in either a pulsed or cw output mode. Organic dye molecules are dissolved in methanol or other liquid solvents, and they have very complex level structures, in which the closely spaced rovibronic levels of the ground and excited states are collision-broadened, forming indistinguishable sublevels, due to the interaction with the solvent, which is beneficial for tuning continuously across a range of wavelengths from $\sim$300–1,100 nm when different dyes are used. The general level structure and the transitions related to the dye laser action are shown in Fig. 4.18.

Up to now, there are hundreds of different dyes developed for active laser media. Under excitation by a suitable laser or flashlamp light with wavelength at the visible or ultraviolet range, dye

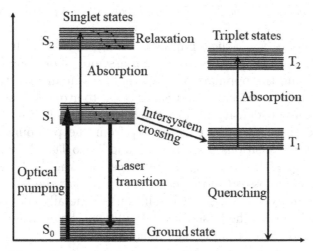

Fig. 4.18.   Level and transition diagram for a dye laser.

molecules in a liquid solvent are normally excited to the singlet state S_1 from the ground state S_0, which are thermally populated with the Boltzmann distribution. The molecules in the excited sublevels of S_1 will decay non-radiatively to the lowest sublevel of the S_1 state quickly in a typical ps time scale. The molecules in the lowest sublevel of the S_1 state relax either radiatively or non-radiatively to the ground state S_0 with a lifetime of about a few ns. It is also possible that the molecules in the lowest sublevel of the S_1 state are excited to the higher singlet state S_2, or transferred to the triplet state T_1 through radiationless transition (intersystem crossing).

At sufficiently strong laser pumping, population inversion between the lowest sublevel in S_1 and higher rovibronic levels in S_0, which normally have a small or negligible population at room temperature, is achieved. In a cavity, laser action will occur. Meanwhile, the stimulated emission populates the higher rovibronic levels in S_0, which quickly relaxes to the lower rovibronic levels by collisions with the solvent molecules. Therefore, the dye laser system can be categorized into a 4-level system. Moreover, the transitions from S_1 to different higher rovibronic levels of S_0 cover a broad spectral range. Therefore, a frequency-selective component, normally a grating, is normally installed in the cavity to reduce the laser linewidth, and by tuning the grating or cavity end mirror, a continuous tunability of laser wavelengths within the fluorescence band of the dye can be obtained.

In dye lasers, when the pumping is strong enough, the intersystem crossing transitions from S_1 to T_1 can assemble more population in T_1, decreasing the population in S_1, thereby diminishing the population inversion between S_1 and S_0. In order to avoid the undesirably large population density in T_1, which has a long lifetime, one has to remove these dye molecules in T_1 from the pumping zone. In addition, the transitions from S_1 to S_2 and T_1 to T_2, which may partially overlap with the S_1–S_0 fluorescence spectrum, will also introduce additional losses (fluorescence quenching) for the lasing action. Therefore, in dye lasers, the dye solution is generally contained in a flowing cell so that the fresh dye molecules can be optically pumped for population inversion between S_1 and S_0.

4.13   Closing Remark

This chapter intends to give an elementary physical account of the operation of a laser oscillator. It is not meant to be complete. Interested readers should consult specialized books devoted entirely to laser operations.

References

A. Einstein (1917). The quantum theory of radiation. *Physikalische Zeitschrift*, *18*, 121.

Chapter 5

Paraxial Gaussian Wave Propagation and Modes

The real laser beam coming out of a resonator does not have uniform intensity across it. Different definite intensity distributions across the beam can be created by controlling the geometrical parameters of the laser cavity. Each distribution corresponds to the solution of the wave equation satisfying the appropriate boundary conditions defined by the geometrical parameters. The most popular laser beam is a cylindrical beam emitted by a laser resonator bounded by two circular mirrors whose g parameters satisfy the stability conditions (Chapter 4). Each solution of the wave equation is called a mode of the laser resonator. For pedagogical reasons, we start by considering the simplest transverse mode, which is a Gaussian spherical wave and its propagation, and then come back to the discussion of cavity modes. The discussion on the propagation of a Gaussian spherical laser beam will lead us at the same time, to the understanding of focusing and collimating such laser beams and designing spatial filters, etc.

5.1 Definition: Spherical Wave

It is a propagating wave whose phase (or wavefront) in a complex amplitude description is a spherical surface emanating from a point source in space and whose amplitude decreases inversely as the

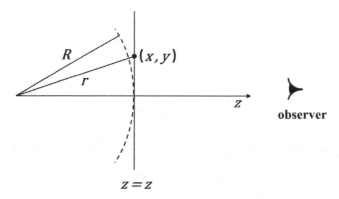

Fig. 5.1.   Paraxial approximation of a spherical wavefront.

distance from the source,

$$\text{i.e.,} \quad \text{Spherical Wave} \propto \frac{f(kr \pm \omega t)}{r} = \frac{e^{-i(kr \pm \omega t)}}{r}, \tag{5.1}$$

where r is the radius of the spherical wave. $\frac{1}{r}$ represents the attenuation of the wave. In spherical coordinates (r, θ, ϕ) with the point source as origin, $f(kr \pm \omega t)$ represents the propagating wave function either diverging from the point source $[f(kr - \omega t)]$ or converging towards the point source $[f(kr + \omega t)]$. It depends only on r (the radius) and independent direction (see Sections 1–3).

We now consider the wavefront in the vicinity of the direction of observation, i.e., the z-direction, as shown in Fig. 5.1. We assume a paraxial laser beam whose wavefront is a spherical diverging wave. We look at its propagation in the vicinity of the z-direction. Hence, at time t, when the wavefront reaches the $(x - y)$ plane at position z,

$$\psi(r) \sim \frac{e^{-i(kr - \omega t)}}{r}, \tag{5.2}$$

$$\sim \frac{e^{-ikr}}{r}, \quad \text{at a fixed time } t. \tag{5.3}$$

We now ask what the wave function is at a point (x, y, z) on the plane $z = z$. We have

$$r^2 = x^2 + y^2 + z^2. \tag{5.4}$$

Since we are concerned only with paraxial rays, $x \ll z, y \ll z$, and Eq. (5.4) becomes

$$r^2 = z^2 \left(1 + \frac{x^2}{z^2} + \frac{y^2}{z^2} \right), \tag{5.5}$$

$$r = z \left[1 + \frac{x^2}{z^2} + \frac{y^2}{z^2} \right]^{1/2},$$

$$\simeq z \left[1 + \frac{x^2}{2z^2} + \frac{y^2}{2z^2} \right], \tag{5.6}$$

where we have expanded the square root in a Taylor series and kept only the first two terms because $x \ll z, y \ll z$. Equation (5.3) becomes:

$$\psi(r) \sim \frac{e^{-ikz\left[1+\frac{x^2}{2z^2}+\frac{y^2}{2z^2}\right]}}{z\left[1 + \frac{x^2}{2z^2} + \frac{y^2}{2z^2}\right]}. \tag{5.7}$$

This is the part of the wavefront propagation in the direction $r = (x^2 + y^2 + z^2)^{1/2}$, *not* in the z-direction. However, because of the assumption of a paraxial ray, the transverse extent of the wavefront is small ($x \ll z, y \ll z$) and around the z-axis. We thus make the approximation that r is almost in the z-direction so that $r = z = R$, where R is the radius of curvature of the wavefront at the $z = z$ plane. Hence, Eq. (5.7) becomes:

$$\psi \sim \frac{e^{-ikR\left[1+\frac{x^2}{2R^2}+\frac{y^2}{2R^2}\right]}}{R\left[1 + \frac{x^2}{2R^2} + \frac{y^2}{2R^2}\right]}. \tag{5.8}$$

The nominator in Eq. (5.8),

$$\text{i.e.,} \quad e^{-ikR\left[1+\frac{x^2}{2R^2}+\frac{y^2}{2R^2}\right]}, \tag{5.9}$$

represents an approximated propagating wavefront of a paraxial spherical wave at the point (x, y) on the plane $z = z$. The denominator is simply an attenuation factor. The transverse phase variation of Eq. (5.9) is

$$e^{-ikR(x^2+y^2)/2R^2} = e^{-ik(x^2+y^2)/2R}. \tag{5.10}$$

5.2 Definition: Gaussian Amplitude Variation of a Wavefront

A general wavefront on the plane $z = z$ may vary in amplitude from point to point on the plane. In the present case, we assume the variation follows a Gaussian distribution,

$$\text{i.e.,} \quad \text{Wave amplitude} \sim e^{-(x^2+y^2)/w^2}, \tag{5.11}$$

where $w \equiv$ *spot size* of the Gaussian distribution.

5.3 Definition: Gaussian Spherical Laser Beam

Using the definitions in the previous two sections, a paraxial laser beam having a spherical wavefront and a Gaussian amplitude distribution across a plane at $z = z$ can thus be represented by combining Eqs. (5.10) and (5.11). We define the transverse part of the wavefront as:

$$u(x, y) = \sqrt{\frac{2}{\pi}} \frac{1}{w} e^{-ikR\left(\frac{x^2}{2R^2} + \frac{y^2}{2R^2}\right)} e^{-(x^2+y^2)/w^2}. \tag{5.12}$$

The factor $\sqrt{\frac{2}{\pi}}\frac{1}{w}$ is a normalizing factor so that $u(x,y)$ represents the electric field strength of the laser beam.

$$\int |u(x, y)|^2 dx dy = 1. \tag{5.13}$$

Equation (5.12) can be made more compact by grouping the exponentials together:

$$-ikR\left(\frac{x^2 + y^2}{2R^2}\right) - \frac{x^2 + y^2}{w^2}$$

$$= \left[-i\frac{2\pi}{\lambda}R \cdot \frac{1}{2R^2} - \frac{1}{w^2}\right](x^2 + y^2),$$

$$= -i\frac{\pi}{\lambda}\left[\frac{1}{R} - \frac{i\lambda}{\pi w^2}\right](x^2 + y^2),$$

$$\equiv -i\frac{\pi}{\lambda}\frac{x^2 + y^2}{q}, \tag{5.14}$$

where

$$\frac{1}{q} \equiv \frac{1}{R} - i\frac{\lambda}{\pi w^2} \tag{5.15}$$

and λ is the wavelength.

Substituting Eq. (5.14) into (5.12):

$$u(x,y) = \sqrt{\frac{2}{\pi}}\frac{1}{w} \cdot e^{-i\frac{\pi}{\lambda}\frac{x^2+y^2}{q}}. \tag{5.16}$$

This can be recast in a similar form as Eq. (5.10),

$$\text{i.e.,} \quad u(x,y) = \sqrt{\frac{2}{\pi}}\frac{1}{w}e^{-ik\frac{x^2+y^2}{2q}}. \tag{5.17}$$

Comparing Eqs. (5.17) and (5.10), we can say that q is a complex radius of curvature of the Gaussian spherical laser beam.

5.4 Huygen–Fresnel's Diffraction Approach to the Propagation of a Ygavefront

We shall make use of the consequence of the Huygen–Fresnel diffraction and examine how a given wavefront propagates through a homogeneous isotropic linear medium. The medium is usually characterized by the index of refraction. In the present discussion, the index n is incorporated in the notation of the wavelength λ such that:

$$\lambda_0 = n\lambda, \tag{5.18}$$

where λ_0 is the wavelength in vacuum. (Equation (5.18) can be verified by noting that the frequency v of the electromagnetic (EM) wave is the same in both the vacuum and homogeneous isotropic linear medium, so that $v = c/\lambda_0 = v/\lambda$, where v is the speed of light in the medium. Since $n = c/v$, we obtain Eq. (5.18).) Thus, all formulations will look as if the propagation is in the "vacuum" with wavelength λ.

We consider a wavefront at the plane $z = z_0$. The transverse part of the wavefront is $u_0(x_0, y_0)$. Referring to Fig. 5.2, we would like to know what this wavefront becomes after propagating to the plane $z = z'$. According to Huygen–Fresnel's principle of superposition

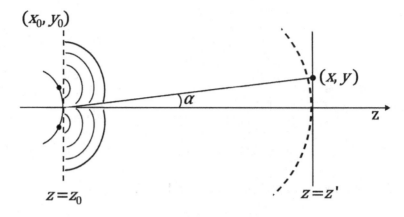

input plant

Fig. 5.2. Diffraction of a spherical paraxial wavefront after propagation through a distance in space.

of secondary waves, each point on the wavefront at $z = z_0$ can be considered as a point source of secondary spherical waves so that the wavefront at $z = z'$ could be considered as the superposition of these secondary waves. The result of such a consideration gives (see any optics text on diffraction):

$$u(x,y,z) = \frac{i}{\lambda} \int\!\!\int_{\text{input}} u_0(x_0,y_0) \frac{1 + \cos\alpha}{2} \frac{e^{-ik|r-r_0|}}{|r - r_0|} dx_0 dy_0, \quad (5.19)$$

where

$$r_0 \equiv (x_0, y_0, z_0),$$
$$r = (x, y, z').$$

Equation (5.19) is valid under the following assumptions:

(a) All dimensions $\gg \lambda$
(b) $\Delta u(x,y) \equiv$ variation of $u(x,y) \ll \lambda$
(c) $\Delta u_0(x_0, y_0) \ll \lambda$
   If we make some further assumptions:
(d) The wavefront is that of a paraxial ray,

$$\text{i.e.,} \quad \alpha \approx 0 \quad \text{and} \quad x, x_0 \ll z' - z_0,$$
$$y, y_0 \ll z' - z_0.$$

These lead to:

$$\frac{1 + \cos\alpha}{2} \approx 1 \tag{5.20}$$

and $\quad |\boldsymbol{r} - \boldsymbol{r}_0| \approx (z' - z_0) \equiv z \tag{5.21}$

and $\quad k|\boldsymbol{r} - \boldsymbol{r}_0| = k[(z - z_0)^2 + (y - y_0)^2 + (x - x_0)^2]^{1/2}$

$$\approx kz\left[1 + \frac{(x - x_O)^2}{2z^2} + \frac{(y - y_O)^2}{2z^2}\right] \tag{5.22}$$

(see Eq. (5.6)).

Inspecting Eq. (5.19), we see that it represents a superposition of spherical waves

$$\frac{e^{-ik|\boldsymbol{r} - \boldsymbol{r}_0|}}{|\boldsymbol{r} - \boldsymbol{r}_0|}$$

with the appropriate weighting factor $u_o(x_o, y_o)$ and a geometrical factor $\frac{1 + \cos\alpha}{2}$. Using Eqs. (5.20) to (5.22), Eq. (5.19) becomes

$$u(x, y, z) = \frac{i}{\lambda} \iint_{input\,plane} u_0(x_0, y_0) \frac{e^{-ikz\left[1 + \frac{(x - x_0)^2}{2z^2} + \frac{(y - y_0)^2}{2z^2}\right]}}{z} dx_0 dy_0,$$

$$= \frac{ie^{-ikz}}{\lambda z} \iint_{input\,plane} u_0(x_0, y_0) e^{-ik\left[\frac{(x - x_0)^2}{2z} + \frac{(y - y_0)^2}{2z}\right]} dx_0 dy_0 . \tag{5.23}$$

Equation (5.23) is the general integral equation of the propagation of a paraxial wavefront a long distance z away.

5.5 Propagation of a Gaussian Plane Wave

We start to apply Eq. (5.23) to different special cases. The present one is a *Gaussian plane wave*. According to Eq. (5.3), a spherical wave becomes a plane wave when $r \to \infty$. This would mean that the spherical surface becomes flatter and flatter, and the center of curvature recedes to infinity. Hence, for a Gaussian plane wave, using

Eq. (5.16) and Eq. (5.15) with $R \to \infty$, we have

$$u_0(x_0, y_0) = \sqrt{\frac{2}{\pi}} \frac{1}{w_0} e^{-i\frac{\pi}{\lambda} \frac{x_0^2 + y_0^2}{q_0}}, \tag{5.24}$$

where

$$q_0^{-1} = 0 - i\frac{\lambda}{\pi w_0^2}, \quad \text{(from Eq. (5.15))},$$

$$q_0 = \frac{\pi w_0^2}{-i\lambda} = \frac{i\pi w_0^2}{\lambda}, \tag{5.25}$$

where w_0 is the spot size at (x_0, y_0). Substituting into Eq. (5.24), we have

$$u_0(x_0, y_0) = \sqrt{\frac{2}{\pi}} \frac{1}{w_0} e^{-i\frac{\pi}{\lambda} \cdot \frac{x_0^2 + y_0^2}{\pi w_0^2/(-i\lambda)}},$$

$$= \sqrt{\frac{2}{\pi}} \frac{1}{w_0} e^{-\frac{x_0^2 + y_0^2}{w_0^2}}, \tag{5.26}$$

which is a Gaussian distribution. This spot size of a Gaussian plane wave is called the beam waist. Substituting Eq. (5.26) into (5.23), we get

$$u(x, y, z) = \frac{ie^{-ikz}}{\lambda z} \iint_{-\infty}^{\infty} \sqrt{\frac{2}{\pi}} \frac{1}{w_0} e^{-\frac{x_0^2 + y_0^2}{w_0^2}} e^{-ik\left[\frac{(x-x_0)^2}{2z} + \frac{(y-y_0)^2}{2z}\right]} dx_0 dy_0. \tag{5.26'}$$

Simplifying and integrating, we obtain

$$u(x, y, z) = \sqrt{\frac{2}{\pi}} \frac{1}{w(z)} \exp\left\{-i[(kz - \psi(z))]\right\} \exp\left[-i\frac{k}{2} \frac{x^2 + y^2}{q(z)}\right], \tag{5.27}$$

where

$$q(z) \equiv q_0 + z = z + i\frac{\pi w_0^2}{\lambda}, \tag{5.28}$$

$$\frac{1}{q(z)} = \frac{1}{R(z)} - i\frac{\lambda}{\pi w^2(z)} = \frac{1}{z + i\pi w_0^2/\lambda}, \tag{5.29}$$

$$w(z) = w_0 \sqrt{1 + \left(\frac{\lambda z}{\pi w_0^2}\right)^2}, \tag{5.30}$$

$$\psi(z) = \tan^{-1}\left(\frac{\lambda z}{\pi w_0^2}\right), \tag{5.31}$$

$$R(z) = z + \left(\frac{\pi w_0^2}{\lambda}\right)^2 \frac{1}{z}. \tag{5.32}$$

For the sake of comparison, we rewrite Eq. (5.24) as follows:

$$u_0(x_0, y_0) = \sqrt{\frac{2}{\pi}} \frac{1}{w_0} \exp\left\{-i\frac{\pi}{\lambda}\frac{x_0^2 + y_0^2}{q_0}\right\} \tag{5.33}$$

and compare with Eq. (5.27). One sees that after propagation through a long distance $z(z \gg (x_0, y_0) \gg \lambda)$, a Gaussian plane wave at the plane $z = z_0$, given by Eq. (5.33), with spot size (waist) w_0 and complex radius of curvature q_0 (Eq. (5.26)) is transformed into a diverging Gaussian spherical wavefront $u(x, y, z)$ due to diffraction, whose spot size is $w(z)$ (Eq. (5.30)), complex radius of curvature $q(z)$ (Eq. (5.28)) and the real radius of curvature $R(z)$ (Eq. (5.32)), plus a phase increase of $[kz - \psi(z)]$ (Eq. (5.31)). This is a very important result because it says that any Gaussian plane wave, after propagation and diffraction, because a diverging Gaussian spherical wave. This means that (see below) the plane $z = z_0$ is the origin of the diverging wave, i.e., it is the "focal plane" of the diverging wave. Since this kind of propagation is reversible, a converging Gaussian spherical wavefront will become a Gaussian plane wave at the focal plane. These are shown in Fig. 5.3.

Consequence (1)

At large $z = z' - z_0$, Eq. (5.32) becomes

$$R(z) = z + \left(\frac{\pi w_0^2}{\lambda}\right)^2 \frac{1}{z},$$

$$\approx z + 0 \quad \left(\text{if } z \gg \frac{\pi w_0^2}{\lambda}\right), \tag{5.34}$$

$$= z.$$

This shows that the real radius of curvature R of the diverging spherical wavefront at $z = z'$ is equal to z. This means that the plane at

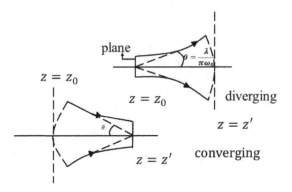

Fig. 5.3.   Converging and diverging Gaussian spherical wave propagation.

$z = z_0$ is indeed at the center of curvature of the outgoing spherical wave at a long distance, i.e., far field (Fig. 5.3).

Consequence (2)

Again, for large $z \gg \frac{\pi w_0^2}{\lambda}$:

$$\omega(z) = \omega_0 \sqrt{1 + \left(\frac{\lambda z}{\pi w_0^2}\right)^2}, \quad \text{(from Eq. (5.30)),}$$

$$\approx \omega_0 \frac{\lambda z}{\pi w_0^2},$$

$$= z \cdot \frac{\lambda}{\pi w_0}, \tag{5.35}$$

$$\equiv z\theta, \tag{5.36}$$

where

$$\theta \equiv \frac{\lambda}{\pi w_0} \tag{5.37}$$

is the (cone) angle of divergence (or simply divergence) at the far field (Fig. 5.3). Thus, looking at the problem in reverse, a converging Gaussian spherical wavefront will converge at the focal plane with a spot size (waist) given by:

$$w(z) = z\theta$$

or

$$w_0(z) \equiv f\theta, \tag{5.38}$$

where f is the focal distance, i.e., the distance between the spherical surface and its focus. Equation (5.38) gives a convenient general estimate of the spot size at the focus of a lens (see below).

5.6 Propagation of a General Gaussian Spherical Wavefront

From Eq. (5.17), the transverse part of the general Gaussian spherical wavefront is:

$$u(x, y) = \sqrt{\frac{2}{\pi}} \frac{1}{w} e^{-k \frac{x^2 + y^2}{2q}}. \quad \text{(Eq. (5.17))}$$

One can substitute this into Eq. (5.23) to obtain the wavefront at a distance z away. However, there is a better and physically clearer way to obtain the result.

Referring back to the previous section, any diverging Gaussian spherical wavefront must have come from a focal plane where the wavefront is a Gaussian plane wave. Similarly, any converging Gaussian spherical wavefront will converge to a focal plane where the wavefront is a Gaussian plane wave. Thus, if we start with a general diverging Gaussian spherical wavefront, we should first ask where its focal plane was and then work from the focal plane, and calculate what it should be at a distance z from the initial Gaussian spherical wavefront. The case of a converging wavefront is similar. Let us now do a calculation for a diverging wavefront at $z = z_1$ from the beam waist at $z = 0$. The initial $u(x,y,z_1)$ is given by Eq. (5.27) with $z = z_1$. We need to find what z_1 is (Fig. 5.4). The complex radius of curvature is, from Eq. (5.29):

$$\frac{1}{q(z_1)} = \frac{1}{R(z_1)} - i \frac{\lambda}{\pi w^2(z_1)}$$

or

$$q(z_1) = \left[\frac{1}{R(z_1)} - i \frac{\lambda}{\pi w^2(z_1)} \right]^{-1}, \quad (5.39)$$

but

$$q(z_1) = q_0 + z_1 = i \frac{\pi w_0^2}{\lambda} + z_1 \quad \text{(from Eq. (5.28))}. \quad (5.40)$$

Equating the real and imaginary parts of the right hand sides of Eqs. (5.39) and (5.40), we have:

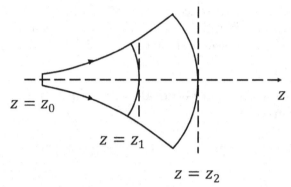

$$z = z_0$$

$$z = z_1$$

$$z = z_2$$

Fig. 5.4. Determination of the wavefront of a diverging Gaussian spherical wave after propagation through a distance in space.

Real part:

$$z_1 = Re\left[\frac{1}{R(z_1)} - i\frac{\lambda}{\pi w^2(z_1)}\right]^{-1},$$

$$= \frac{\frac{1}{R(z_1)}}{1/R^2(z_1) + \frac{\lambda^2}{\pi^2 w^4(z_1)}}. \tag{5.41}$$

Imaginary part:

$$\frac{\pi w_0^2}{\lambda} = \frac{\frac{\lambda}{\pi w^2(z_1)}}{1/R^2(z_1) + \frac{\lambda^2}{\pi^2 w^4(z_1)}}$$

or

$$w_0 = \frac{\lambda/[\pi w(z_1)]}{\left\{1/R^2(z_1) + \frac{\lambda^2}{\pi^2 w^4(z_1)}\right\}^{1/2}}. \tag{5.42}$$

Let

$$w(z_1) \equiv w_1,$$

$$R(z_1) \equiv R_1.$$

Equation (5.42) becomes

$$w_0 = \frac{w_1}{\left\{1 + \frac{\pi^2 w_1^4}{\lambda^2 R_1^2}\right\}^{1/2}}. \tag{5.43}$$

Now knowing w_0, we find the wavefront at $z = z_2$ from the beam waist (Fig. 5.4). Using Eq. (5.28), the complex radius of curvature is

$$q(z_2) = q_0 + z_2 = z_2 + i\frac{\pi w_0^2}{\lambda}. \tag{5.44}$$

But from

$$\text{Eq. (5.40),} \quad i\frac{\pi w_0^2}{\lambda} = q(z_1) - z_1. \tag{5.45}$$

Substituting Eq. (5.45) into (5.44), we have

$$\begin{aligned} q(z_2) &= z_2 + q(z_1) - z_1, \\ &= q(z_1) + (z_2 - z_1) \end{aligned} \tag{5.46}$$

and the wavefront at $z = z_2$ is given by Eq. (5.27), with $z \to z_2$. We note here that Eq. (5.46) gives the transformation of $q(z_1)$ into $q(z_2)$ for the propagation of a Gaussian spherical wave in a homogeneous medium. In fact, if we do not need to know w_0 explicitly, we can simply bypass Eq. (5.41) to (5.43) and still obtain the main result. The case of a converging wavefront is left as a straightforward exercise for the reader.

5.7 Propagation of a Gaussian Spherical Wavefront Through a Thin Lens

In geometrical optics, a diverging spherical wavefront is transformed into a converging spherical wavefront by thin spherical lens, see Fig. 5.5. We can use the matrix transformation of a thin lens in Chapter 4 to calculate R_2. Note that we are always dealing with paraxial rays.

$$\begin{pmatrix} r_o \\ r_o' \end{pmatrix} = \begin{pmatrix} 1 & 0 \\ -\dfrac{1}{f} & 1 \end{pmatrix} \begin{pmatrix} r_i \\ r_i' \end{pmatrix}, \quad \text{(from Eq.(4.3)),}$$

i.e., $\quad r_o = r_i.$ \hfill (5.47)

and

$$r_o' = r_i' - r_i/f. \tag{5.48}$$

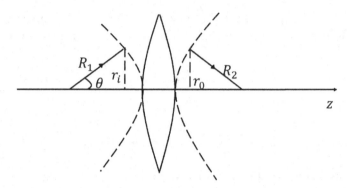

Fig. 5.5.   Propagation of a Gaussian spherical wave through a thin lens.

For paraxial rays, $r'_o \approx r_o/R_2$ and

$$r'_i \approx r_i/R_1; \quad \text{Eq. (5.48)} \Rightarrow \frac{r_o}{R_2} = \frac{r_i}{R_1} - \frac{r_i}{f}.$$

Using Eq. (5.47), this becomes:

$$\frac{1}{R_2} = \frac{1}{R_1} - \frac{1}{f}. \tag{5.49}$$

Now, if the incident wavefront is a Gaussian spherical type, the wavefront is still spherical, while the amplitude distribution is Gaussian. Assuming no loss through the lens, the amplitude distribution at the output side of the lens should be unchanged.

Hence, adding $\left(-\frac{i\lambda}{\pi w_2^2}\right)$ to both sides of Eq. (5.49) gives:

$$\frac{1}{R_2} - i\frac{\lambda}{\pi w_2^2} = \frac{1}{R_1} - i\frac{\lambda}{\pi w_2^2} - \frac{1}{f}. \tag{5.50}$$

But, $w_2 = w_1$ because the amplitude distribution is unchanged for a thin lens. We thus have, using Eq. (5.29):

$$\frac{1}{q_2} = \frac{1}{q_1} - \frac{1}{f}, \tag{5.51}$$

i.e., a thin lens transforms a Gaussian spherical wave in a similar way as it transforms an ordinary spherical wave.

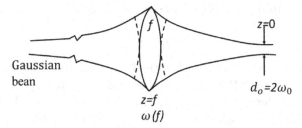

Fig. 5.6. Gaussian spherical wave focusing by a lens.

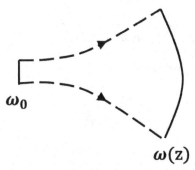

Fig. 5.7. Gaussian spherical wave.

5.8 Focal Spot Size

We can now make use of the result of Section 5.7 to calculate the transverse diameter d_0 at the focus of a thin lens when a Gaussian spherical wave is focused down to a "spot". This is shown in Fig. 5.6. We already know that in the propagation of a Gaussian plane wave into a Gaussian spherical wavefront (Fig. 5.7), the beam radius at z, i.e., $w(z)$, is related to the beam waist w_0 by Eq. (5.30),

$$\text{i.e.,} \quad w(z) = w_0 \left[1 + \left(\frac{z}{z_R} \right)^2 \right]^{\frac{1}{2}}, \quad z_R \equiv \frac{\pi w_0^2}{\lambda},$$

$$\simeq \frac{w_0 z}{z_R}, \quad \text{if } z \gg z_R,$$

$$= \frac{\lambda z}{\pi w_0}.$$

Hence, regrouping the left and right hand sides:

$$w_0 w(z) = \frac{\lambda z}{\pi}. \tag{5.52}$$

Because of the reversibility of optical rays (waves), the propagation in the reverse direction is also true, and Eq. (5.52) is still valid. That is, if the lens transforms the incident Gaussian spherical wave into one that has a waist $w(z)(z = f)$ (see Fig. 5.6) at the output side of the lens, its relationship with w_0 at the focus $(z = 0)$ is also

$$w_0 w(z) \simeq \frac{\lambda z}{\pi},$$

$$w_0 w(f) \simeq \frac{\lambda f}{\pi}.$$

This equation is valid if $f \gg \pi w_0^2/\lambda$. (See Section 5.5, Consequence 2.)

$$w_0 = \frac{\lambda f}{\pi w(f)}$$

or

$$d_0 = 2w_0 \simeq \frac{2\lambda f}{\pi w(f)}. \tag{5.53}$$

Definition: 99% criterion

It just happens that for a Gaussian spherical laser beam, 99% of its power (assuming a cw laser) is concentrated within the diameter (exercise).

$$D = \pi w(f). \tag{5.54}$$

Using this criterion, Eq. (5.53) becomes:

$$d_0 = \frac{2\lambda f}{D}, \tag{5.55}$$

$$= 2\lambda f^\#, \tag{5.56}$$

where

$$f^\# \equiv f/D. \tag{5.57}$$

Equation (5.55) says that if we can measure experimentally at the input side of the thin lens the diameter D of the laser beam containing 99% of its power, and if f and λ are known, we can calculate the focal diameter d_0, which 86% of the laser power passes through.

It should be noted that even if the laser beam is a purely Gaussian spherical wave and is paraxial, the calculation of d_0 is still not exact because we have neglected the aberration of the lens. Moreover, most laser beams are not perfectly Gaussian spherical. Hence, Eqs. (5.53) to (5.57) represent only some practical estimates. Any better knowledge of the real d_0 has to be measured directly at the focus. This poses a challenge in the measurement of the focal spot size of very intense laser pulses. Direct measurement is impossible because any measuring device (film, detector arrays, translating pin-holes, etc.) set at the focus will be severely damaged. Using the beam-splitting technique requires very good beam splitting optical surfaces so that the wavefront is not disturbed. Even more difficult is the measurement of infrared laser focal diameter because of the lack of sensitive material. Although all these difficulties have been overcome, the price is high and there is still a lot of room for improvement.

Often, experimental measurement of the focal spot size is represented by an integrated energy distribution across the focal diameter d_0, as shown in Fig. 5.8. The diameter or full width at half the maximum value of the distribution is defined as the width, or FWHM.

There is another practical way of estimating the focal spot size. This is by way of geometrical optics (Fig. 5.9). The laser beam divergence angle

$$\theta = 2\theta' \tag{5.58}$$

can be calculated as follows:

$$\frac{d_0/2}{f} = \tan \theta' \approx \theta', \quad \text{for small } \theta'.$$

Hence,

$$d_0 = f \cdot 2\theta'$$

or

$$d_0 = f\theta \tag{5.59}$$

Normally, one can measure the divergence angle at the input side of the lens and apply Eq. (5.59) to calculate CJQ. It should be warned again that such measurements, as well as those using Eqs. (5.53) to (5.57), are estimates. More precise measurements have to be made directly.

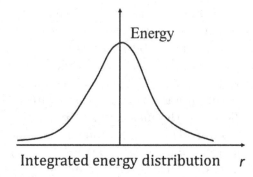

Fig. 5.8.   Energy distribution of a laser beam.

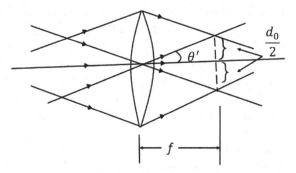

Fig. 5.9.   Quick estimation of the focal spot size.

5.9   Modes

In a laser cavity, the laser radiation "oscillates" at different possible modes. Each oscillating mode is a wave of a fixed frequency with a certain transverse energy (field amplitude) distribution satisfying both the Maxwell equations and the boundary conditions for a stable oscillation, i.e., $0 \leq g_1 g_2 \leq 1$ (see Chapter 4). In practice, one loosely distinguishes these modes into two types, spatial or transverse modes, and axial or temporal modes, although they are interwoven together (see below).

Definition: *Spatial or transverse mode*

At a fixed laser frequency, the cavity can, in general, sustain stable waves with different kinds of transverse amplitude distributions. An example is the Gaussian spherical wave, which is the lowest order

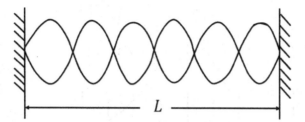

Fig. 5.10.   Standing wave in a laser cavity.

transverse mode in a laser cavity with spherical mirrors. (This mode is sometimes called the fundamental transverse mode.) Higher-order modes can exist whose angle of divergence is larger than that of the fundamental transverse mode. Hence, very often, one puts a diaphragm inside a laser cavity and makes the hole as small as possible so that only the fundamental transverse mode can pass through the hole. The laser output is thus a Gaussian spherical wave. However, the frequency at which this fundamental transverse mode oscillates is not unique. Other allowed axial frequencies can also oscillate with the same transverse distribution. This leads to the following definition.

Definition: *Axial or temporal or longitudinal modes*

At a fixed transverse field distribution in a laser cavity, the stable oscillation demands that there be a standing wave along the axis of the cavity, as shown in Fig. 5.10, i.e., the cavity length L should be:

$$L = n\left(\frac{\lambda}{2}\right), \quad n = \text{very large integer.} \qquad (5.60)$$

Since

$$\lambda = \frac{C}{\nu},$$

where c is the appropriate speed of light in the media inside the cavity and v is the frequency of the laser. Equation (5.60) becomes:

$$L = n \cdot \frac{c}{2\nu}. \qquad (5.61)$$

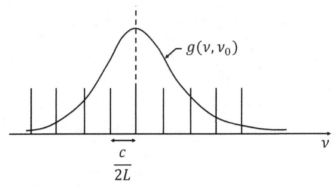

Fig. 5.11.   Gain curve of a laser oscillator and the axial modes of the cavity.

We define ν_n as the frequency corresponding to a large integer n. Equation (5.61) becomes

$$\nu_n = n \left(\frac{c}{2L} \right). \tag{5.62}$$

We call ν_n an axial or longitudinal mode (frequency). In a laser cavity, many such axial modes can be sustained, with the separation between an adjacent pair being:

$$\nu_{n+1} - \nu_n = (n+1) \left(\frac{c}{2L} \right) - n \left(\frac{c}{2L} \right),$$

$$= \frac{c}{2L}, \tag{5.63}$$

which is a constant. Thus, under the gain curve of the laser (Fig. 5.11), there can be several axial modes that experience the gain and hence will oscillate. The simultaneous operation of many axial modes results in mode beating (interference effect) among themselves so that a temporal modulation of the wave exists. If we detect the laser output at a fixed position z along the laser axis, the power will be modulated in time (Fig. 5.12).

5.10   Spatial-Temporal Modes

Any appropriate combination of a frequency and a single transverse amplitude distribution satisfying the lasing conditions in a cavity constitutes a spatial-temporal mode. A general statement of the lasing condition from the point of view of wave propagation is

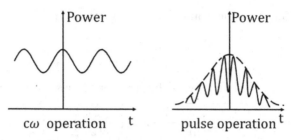

Fig. 5.12. Temporal mode beatings.

the following: The laser field's amplitude distribution and its phase should reproduce themselves after one round trip inside the cavity, i.e., gain = loss and the total phase change after one round trip is $2m\pi$ ($m = 0, 1, 2$). It is the phase change condition that leads to the allowed frequency (mode) oscillation.

For example, for a pedagogical reason, we assume that a plane wave oscillates in the cavity, i.e., apart from the amplitude, the propagating part of the wave is $\exp\{-i(kz - \omega t)\}$ where z is the laser axis. After one round trip, we should have (phase condition: the wave reproduces itself):

$$\exp\{-i(kz - \omega t)\} = \exp\{-i[k(z + 2L) - \omega t]\},$$

$$\exp[-i(k2L)] = 1,$$

$$k2L = 2m\pi \quad (m = 0, 1, 2, \ldots),$$

$$\text{i.e.,} \quad \frac{2\pi}{\lambda} \cdot 2L = 2m\pi,$$

$$L = m\left(\frac{\lambda}{2}\right),$$

which is Eq. (5.60), the condition for stable oscillation.

A more realistic example is a Gaussian spherical wave, which can be a fundamental mode of the cavity. Referring to Eq. (5.27), the phase part due to propagation is $\exp\{-i(kz - \psi(z))\}$ with $\psi(z) = \tan^{-1}\left(\left|\frac{\lambda z}{\pi w_0^2}\right|\right)$. Round trip reproduction of the wave means:

$$\exp\{-i[kz - \psi(z)]\} = \exp\{-i[k(z + 2L) - \psi(z + 2L)]\}$$

or

$$k(z + 2L) - \psi(z + 2L) - (kz - \psi(z)) = 2m\pi,$$

$$\text{i.e.,} \quad k \cdot 2L - [\psi(z + 2L) - \psi(z)] = 2m\pi. \tag{5.64}$$

Equation (5.64) is the condition that governs the laser frequencies through

$$k = \frac{2\pi}{\lambda} = \frac{2\pi\nu}{c}.$$

A general example is a general high order Gaussian mode given by:

$$u_{mn}(x, y, z) = \sqrt{\frac{1}{2^{m+n}m!n!}} \frac{1}{w(z)} H_m\left(\frac{\sqrt{2}x}{w(z)}\right) H_n\left(\frac{\sqrt{2}y}{w(z)}\right)$$

$$\cdot \exp\left(-i\frac{k}{2}\frac{x^2 + y^2}{q(z)}\right) \exp[-ikz + i(m + n + 1)\psi(z)],$$

$$(5.65)$$

where $q(z)$, $\omega(z)$ and $\psi(z)$ are given by Eqs. (5.29) to (5.31), respectively, and where $H_n(x)$ is the Hermite polynomial. This mode becomes the fundamental Gaussian spherical mode when $m = n = 0$, and one often refers to the latter case as a TEM_{∞} mode. Round trip reproduction of wave means:

$$\exp[-ikz + i(m + n + 1)\psi(z)] = \exp[-ik(z + 2L)$$

$$+ i(m + n + 1)\psi(z + 2L)],$$

i.e., $\quad k(z + 2L) - (m + n + 1)\psi(z + 2L)$

$$-[kz - (m + n + 1)\psi(z)] = 2q\pi (q = 0, 1, 2, \ldots)$$

or

$$k \cdot 2L - (m + n + 1)[\psi(z + 2L) - \psi(z)] = 2q\pi. \qquad (5.66)$$

Using

$$k = \frac{2\pi\nu}{c} \equiv \frac{2\pi}{c}\nu_{qmn} \qquad (5.67)$$

And Eqs. (5.29) to (5.31), and after some lengthy calculation (left as an <u>exercise</u> for the reader), one obtains the general expression for the frequency of a general stable cavity Gaussian mode.

$$\nu_{qmn} = \left[q + (m + n + 1)\frac{\cos^{-1}\sqrt{g_1 g_2}}{\pi}\right]\frac{c}{2L}, \qquad (5.68)$$

where

$$g_i = 1 - \frac{L}{R_i}, \quad i = 1, 2$$

and R_i is the radius of curvature of the ith mirror.

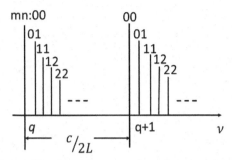

Fig. 5.13. Spatial-temporal modes of a cavity.

In general, on the frequency scale, the modes look like the sketch shown in Fig. 5.13. Because the spacing between the qth and $(q+1)^{\text{th}}$ group of lines is $c/2L$, q labels the axial modes. For each axial mode, there are many transverse modes possible, labeled by (mn). The spacing between adjacent transverse modes depends on $g_1 g_2$.

5.11 Closing Remarks

No attempt is made here to give a detailed description of the modes inside a cavity based on the solution of the Maxwell equations. It is long and tedious and is beyond the scope of this book. Readers who are interested in the details can consult any advanced laser textbook.

Chapter 6

Optical Anisotropy in a Lossless Medium

We now come to a basic physical phenomenon, namely, optical anisotropy, which underlies a large number of passive as well as active optical elements in modern optics. These comprise elements such as electro-optical modulators, acousto-optical modulators, nonlinear optical elements, wave plates (quarter, half, full waves, etc.), Faraday rotators, birefringent filters, etc. Because this basic phenomenon is so important for the understanding of so many modern optical operations, it is necessary to have a thorough discussion of the physical process. The discussion would involve quite a bit of tedious calculation. Quite often, at the end of a set of calculations, the reader is lost in mathematics and has to review what he is up to. As such, for pedagogical reasons, we try to pursue the discussion in a way that is not so systematic in the expert's eyes. The reader will be reminded again and again on what he is looking for and where he is going.

When we send an electromagnetic (EM) wave into a medium, the electric response of the material is fundamentally an induced polarization (dipole moment/volume) generated by the electric field of the EM wave. Assuming an isotropic medium:

$$\boldsymbol{P} = \epsilon_0 \chi \boldsymbol{E},$$

where χ is the susceptibility, and $\boldsymbol{P}$ and $\boldsymbol{E}$ are the polarization and electric field, respectively, in a very small region of the medium.

The usual definition of the field D is:

$$D \equiv P + \epsilon_0 E,$$

$$\equiv \epsilon E,$$

where ϵ is the dielectric constant. It is related to the refractive index, n, and the EM wave's propagation velocity v by:

$$n^2 = \epsilon/\epsilon_0,$$

$$v = \frac{c}{n} = \frac{c}{\sqrt{\epsilon/\epsilon_0}}.$$

Thus, once we know v and E, we can calculate P. Because we usually measure n and/or v, but not P, we would say that the response of the medium is known if we know n and/or v.

In this chapter, the medium is assumed anisotropic. Thus, the induced polarization is also anisotropic. We shall ask what the response of the anisotropic medium to an EM wave is and analyze this response in terms of not P, but the various n's and/or v's, together with the associated fields D's and E's.

6.1 Optical Anisotropy

In the interaction between radiation and matter, the response of the material medium is not always isotropic. That is to say, the response depends on the direction of propagation as well as the polarization of the radiation. We assume that only the electric response of the medium is anisotropic while the magnetic response is always isotropic. This is reasonable (apart from some exceptions, see Section 10.1) even for laser fields strong enough to damage the material because:

$$E = cB, \tag{6.1}$$

where E is the electric field, B is the associated magnetic field of the radiation, and c is the speed of light, i.e., $E \gg B$. (If the laser field is very strong, molecular dissociation, vaporization and plasma formation, etc., will take place. Even though the magnetic field of the laser may now become significant, its effect on the initial material will not be seen anymore because the material is already damaged.)

With this in mind, we consider an electrically anisotropic medium and make the following statement.

The dielectric constant of the anisotropic medium is not a constant any more, but a tensor. This dielectric tensor is symmetric. In mathematical form, it is:

$$\epsilon = \begin{pmatrix} \epsilon_{11} & \epsilon_{12} & \epsilon_{13} \\ \epsilon_{21} & \epsilon_{22} & \epsilon_{23} \\ \epsilon_{31} & \epsilon_{32} & \epsilon_{33} \end{pmatrix} \tag{6.2}$$

and

$$\epsilon_{ij} = \epsilon_{ji}, \quad \text{(symmetric condition)}. \tag{6.3}$$

6.1.1 Confirmation

The dielectric constant describes the response of the material to the electric field of the radiation. Under the influence of an electric field, the medium will be polarized. The induced polarization P (dipole moment per unit volume), the displacement D, and the electric field E are related by:

$$D = \epsilon_0 E + P. \tag{6.4}$$

In an isotropic medium, D, E and P are parallel to one another. However, in an anisotropic medium, they are not. Also

$$P = \epsilon_0 \chi E, \tag{6.5}$$

where χ is the electric susceptibility tensor. In matrix form, Eq. (6.5) becomes

$$\begin{pmatrix} P_1 \\ P_2 \\ P_3 \end{pmatrix} = \epsilon_0 \begin{pmatrix} \chi_{11} & \chi_{12} & \chi_{13} \\ \chi_{21} & \chi_{22} & \chi_{23} \\ \chi_{31} & \chi_{32} & \chi_{33} \end{pmatrix} \begin{pmatrix} E_1 \\ E_2 \\ E_3 \end{pmatrix}. \tag{6.6}$$

Substituting into Eq. (6.4), we have

$$\begin{pmatrix} D_1 \\ D_2 \\ D_3 \end{pmatrix} = \epsilon_0 \begin{pmatrix} E_1 \\ E_2 \\ E_3 \end{pmatrix} + \epsilon_0 \begin{pmatrix} \chi_{11} & \chi_{12} & \chi_{13} \\ \chi_{21} & \chi_{22} & \chi_{23} \\ \chi_{31} & \chi_{32} & \chi_{33} \end{pmatrix} \begin{pmatrix} E_1 \\ E_2 \\ E_3 \end{pmatrix}$$

$$= \epsilon_0 \begin{pmatrix} 1+\chi_{11} & \chi_{12} & \chi_{13} \\ \chi_{21} & 1+\chi_{22} & \chi_{23} \\ \chi_{31} & \chi_{32} & 1+\chi_{33} \end{pmatrix} \begin{pmatrix} E_1 \\ E_2 \\ E_3 \end{pmatrix}. \tag{6.7}$$

Definition: dielectric tensor ϵ

$$\epsilon \equiv \epsilon_0 \begin{pmatrix} 1 + \chi_{11} & \chi_{12} & \chi_{13} \\ \chi_{21} & 1 + \chi_{22} & \chi_{23} \\ \chi_{31} & \chi_{32} & 1 + \chi_{33} \end{pmatrix}, \tag{6.8}$$

$$\equiv \begin{pmatrix} \epsilon_{11} & \epsilon_{12} & \epsilon_{13} \\ \epsilon_{21} & \epsilon_{22} & \epsilon_{23} \\ \epsilon_{31} & \epsilon_{32} & \epsilon_{33} \end{pmatrix}. \tag{6.9}$$

Substituting Eqs. (6.8) and (6.9) into (6.7):

$$\begin{pmatrix} D_1 \\ D_2 \\ D_3 \end{pmatrix} = \begin{pmatrix} \epsilon_{11} & \epsilon_{12} & \epsilon_{13} \\ \epsilon_{21} & \epsilon_{22} & \epsilon_{23} \\ \epsilon_{31} & \epsilon_{32} & \epsilon_{33} \end{pmatrix} \begin{pmatrix} E_1 \\ E_2 \\ E_3 \end{pmatrix} \tag{6.10}$$

or

$$D_i = \sum_{i,j=1}^{3} \epsilon_{ij} E_j. \tag{6.11}$$

This explains Eq. (6.2). We still need to prove $\epsilon_{ij} = \epsilon_{ji}$.

Assumptions:

(a) ϵ_{ij} $(i,j = 1,2,3)$ are real. (Complex values of ϵ_{ij} are possible. This would lead to complex indices of refraction resulting in the absorption by the medium. Thus, this assumption of real ϵ_{ij} is equivalent to assuming a lossless or non-absorbing medium.)
(b) The medium is homogeneous and magnetically isotropic.
(c) The medium is linear, i.e., ϵ_{ij}'s do not depend on $\boldsymbol{E}$.

Now, two of the Maxwell equations in the medium without a current are:

$$\nabla \times \boldsymbol{H} = \frac{\partial \boldsymbol{D}}{\partial t}, \tag{6.12}$$

$$\nabla \times \boldsymbol{E} = -\frac{\partial \boldsymbol{B}}{\partial t}, \tag{6.13}$$

where

$$\boldsymbol{B} = \mu \boldsymbol{H}, \tag{6.14}$$

$$\mu \approx \mu_0.$$

Because of the assumption that the medium is isotropic magnetically, i is a constant real number.

$$\boldsymbol{E} \cdot [\text{Eq. } (6 - 12)] \quad \text{gives}$$

$$\boldsymbol{E} \cdot (\nabla \times \boldsymbol{H}) = \boldsymbol{E} \cdot \left(\frac{\partial \boldsymbol{D}}{\partial t} \right). \tag{6.15}$$

Now,

$$\nabla \cdot (\boldsymbol{E} \times \boldsymbol{H}) = \boldsymbol{H} \cdot (\nabla \times \boldsymbol{E}) - \boldsymbol{E} \cdot (\nabla \times \boldsymbol{H}), \tag{6.16}$$

which is a vector identity. Substituting Eq. (6.16) into (6.15), we have

$$\boldsymbol{H} \cdot (\nabla \times \boldsymbol{E}) - \nabla \cdot (\boldsymbol{E} \times \boldsymbol{H}) = \boldsymbol{E} \cdot \left(\frac{\partial \boldsymbol{D}}{\partial t} \right). \tag{6.17}$$

Since

$$\boldsymbol{H} \cdot (\nabla \times \boldsymbol{E}) = \boldsymbol{H} \cdot \left(-\frac{\partial \boldsymbol{B}}{\partial t} \right), \quad (\text{from Eq. } (6.13)),$$

Eq. (6.17) becomes

$$-\nabla \cdot (\boldsymbol{E} \times \boldsymbol{H}) = \boldsymbol{H} \cdot \left(\frac{\partial \boldsymbol{B}}{\partial t} \right) + \boldsymbol{E} \cdot \left(\frac{\partial \boldsymbol{D}}{\partial t} \right) \tag{6.18}$$

and since $\boldsymbol{S} \equiv$ Poynting vector $\equiv \boldsymbol{E} \times \boldsymbol{H}$,

$$\text{Eq. } (6.18) \text{ becomes } - \nabla \cdot \boldsymbol{s} = \mu \boldsymbol{H} \cdot \left(\frac{\partial \boldsymbol{H}}{\partial t} \right) + \boldsymbol{E} \cdot \left(\frac{\partial \boldsymbol{D}}{\partial t} \right). \tag{6.19}$$

But

$$-\nabla \cdot \boldsymbol{s} = \frac{d\mathrm{W}}{dt} \equiv \text{rate of change of EM energy density,} \tag{6.20}$$

which is simply a continuity equation for energy flow, i.e., the rate of change of electromagnetic energy density is equal to the outflow of energy flux $(-\nabla \cdot \boldsymbol{s})$.

Since

$$W = W_{\text{electric}} + W_{\text{magnetic}} \equiv W_e + W_m$$

$$= \frac{1}{2} \boldsymbol{E} \cdot \boldsymbol{D} + \frac{1}{2} \boldsymbol{H} \cdot \boldsymbol{B}. \tag{6.21}$$

Note that $\boldsymbol{E}$ is not parallel to $\boldsymbol{D}$ while $\boldsymbol{B} = \mu \boldsymbol{H} (\mu = \text{constant})$ i.e., $\boldsymbol{B}$ is parallel to $\boldsymbol{H}$.

From Eq. (6.21):

$$\therefore \frac{dW}{dt} = \frac{d}{dt}\left\{\frac{1}{2}\sum_{i,j=1}^{3} E_i \epsilon_{ij} E_j\right\} + \frac{1}{2}\frac{d\boldsymbol{H}}{dt}\cdot(\mu\boldsymbol{H}) + \frac{1}{2}\boldsymbol{H}\cdot\frac{d}{dt}(\mu\boldsymbol{H}),$$

$$\frac{dW}{dt} = \frac{1}{2}\left[\sum_{i,j}\dot{E}_i\epsilon_{ij}E_j + \sum_{i,j}E_i\epsilon_{ij}\dot{E}_j\right] + \mu\boldsymbol{H}\cdot\left(\frac{d\boldsymbol{H}}{dt}\right). \qquad (6.22)$$

Comparing Eqs. (6.19), (6.20) and (6.22), we have, by equating the right hand side of Eq. (6.19), and that of Eq. (6.22):

$$\frac{1}{2}\left[\sum_{ij}\dot{E}_i\epsilon_{ij}E_j + \sum_{ij}E_i\epsilon_{ij}\dot{E}_j\right] = \boldsymbol{E}\cdot\frac{\partial\boldsymbol{D}}{\partial t}$$

$$= \sum_i E_i\frac{\partial}{\partial t}\left(\sum_j \epsilon_{ij}E_j\right)$$

$$= \sum_{ij} E_i\epsilon_{ij}\dot{E}_j$$

$$\frac{1}{2}\sum_{ij}\dot{E}_i\epsilon_{ij}E_j = \frac{1}{2}\sum_{ij}E_i\epsilon_{ij}\dot{E}_j. \qquad (6.23)$$

Exchanging i and j on the right hand side of Eq. (6.23) does not change its value.

Hence, Eq. (6.23) becomes:

$$\frac{1}{2}\sum_{ij}\dot{E}_i\epsilon_{ij}E_j = \frac{1}{2}\sum_{ij}E_j\epsilon_{ji}\dot{E}_i.$$

Thus,

$$\epsilon_{ij} = \epsilon_{ji}, \qquad (6.24)$$

which proves the symmetric condition of Eq. (6.3).

6.2 Electromagnetic Wave Interaction with an Anisotropic Medium (General Considerations)

We assume that the anisotropic medium is lossless (hence transparent) to the EM wave in question. Thus, ϵ'_{ij}s are real. Let a linearly polarized, monochromatic, plane EM wave propagate in the direction $\hat{k}$ inside the medium. The transverse electric field of the EM wave will induce a polarization $\boldsymbol{P}$ given by Eq. (6.6), which re-radiates.

Statement 1

Given a propagation direction, the re-radiated EM waves consist of two monochromatic plane waves with two different phase·velocities governed by the surface of wave normals, i.e., there is a unique index of refraction for each of the two waves. When $\hat{k}$ is along some specific (generally two) directions (called optic axes), only one secondary wave is re-radiated. These are shown schematically in Fig. 6.1. The phenomenon is called birefringence.

Confirmation

(1) Dielectric tensor

By a suitable transformation, the dielectric tensor can be transformed into a diagonal form (in its so-called *principal coordinate system*, cf.

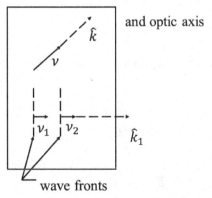

Fig. 6.1. An EM wave propagating in an anisotropic medium in an arbitrary direction $\hat{k}_1$ induces two waves propagating at two different velocities v_1 and v_2, whereas in the direction of the optic axis, only one wave propagates at a velocity v.

any 3D analytical geometry or matrix algebra),

$$\text{i.e.,} \quad \epsilon = \begin{pmatrix} \epsilon_X & 0 & 0 \\ 0 & \epsilon_Y & 0 \\ 0 & 0 & \epsilon_Z \end{pmatrix}. \tag{6.25}$$

Henceforth, we name the $x-y-z$ co-ordinates the *principal dielectric axes*.

(2) Wave representation

We assume that the EM wave is a plane wave interacting with the material inside the anisotropic medium,

$$\text{i.e.,} \quad \boldsymbol{E}(\boldsymbol{r},t) = \boldsymbol{E}\exp[i(\omega t - \boldsymbol{k} \cdot \boldsymbol{r})], \tag{6.26}$$

$$\boldsymbol{H},(\boldsymbol{r},t) = \boldsymbol{H}\exp[i(\omega t - \boldsymbol{k} \cdot \boldsymbol{r})]. \tag{6.27}$$

(3) Wave equation

We rewrite two of the four Maxwell equations:

$$\nabla \times \boldsymbol{E} + \frac{\partial \boldsymbol{B}}{\partial t} = 0, \quad \text{Faraday's law,} \tag{6.28}$$

$$\nabla \times \boldsymbol{H} - \frac{\partial \boldsymbol{D}}{\partial t} = 0, \quad \text{Modified Ampere's law.} \tag{6.29}$$

Substituting Eqs. (6.26) and (6.27) into Eqs. (6.28) and (6.29), and using the following consequences when operating on plane waves:

$$\nabla = -i\boldsymbol{k}, \tag{6.30}$$

$$\frac{\partial}{\partial t} = i\omega \tag{6.31}$$

(*Exercise*: The reader should try to prove Eqs. (6.30) and (6.31)) we have

$$\boldsymbol{k} \times \boldsymbol{E} = \omega\mu\boldsymbol{H}, \tag{6.32}$$

$$\boldsymbol{k} \times \boldsymbol{H} = -\omega\epsilon\boldsymbol{E} \tag{6.33}$$

$\boldsymbol{k} \times$ Eq. (6.32) gives:

$$\boldsymbol{k} \times \boldsymbol{k} \times \boldsymbol{E} = \omega\mu\boldsymbol{k} \times \boldsymbol{H},$$

$$= \omega\mu(-\omega\epsilon)\boldsymbol{E}, \quad (\text{using Eq. (6.33)}),$$

$$\boldsymbol{k} \times \boldsymbol{k} \times \boldsymbol{E} + \omega^2\mu\epsilon\boldsymbol{E} = 0. \tag{6.34}$$

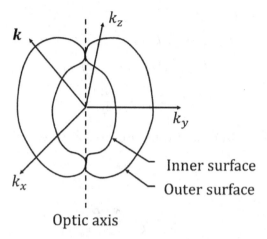

Optic axis

Fig. 6.2.   A cross-section of the surface of wave normals.

Equation (6.34) is the equivalent of the wave equation. Substituting Eq. (6.25) into (6.34), and after some lengthy calculation, one obtains an equation $\omega(\boldsymbol{k}) = $ constant in the k-space (i.e., wave vector space). This equation is called Fresnel's equation of wave normals. In the 3D k-space, it represents two sheets of connected surfaces (see Fig. 6.2), for a constant ω. Along a general direction of propagation $\hat{k}$, the wave vector $\boldsymbol{k}$ will intersect the two surfaces, giving two solutions of k. Since $k = \frac{n\omega}{c} = \frac{\omega}{v}$, two solutions of k means two values of v, the phase velocity, or two indices of refraction. This is what has just been stated.

(4) Fresnel's equation of wave normals

We now demonstrate two methods of calculation that lead to Fresnel's equation of wave normals (i.e., $\omega(\boldsymbol{k}) = $ constant).

Method (a)
Since

$$\boldsymbol{a} \times (\boldsymbol{b} \times \boldsymbol{c}) = (\boldsymbol{a} \cdot \boldsymbol{c})\boldsymbol{b} - (\boldsymbol{a} \cdot \boldsymbol{b})\boldsymbol{c},$$

which is a vector identity, Eq. (6.34) becomes

$$(\boldsymbol{k} \cdot \boldsymbol{E})\boldsymbol{k} - k^2 \boldsymbol{E} + \omega^2 \mu\epsilon \boldsymbol{E} = 0. \qquad (6.35)$$

In component form:

$$(\boldsymbol{k} \cdot \boldsymbol{E})k_i - k^2 E_i + \omega^2 \mu \epsilon_i E_i = 0, \qquad (6.36)$$

$$(i = 1, 2, 3 \quad \text{or} \quad x, y, z)$$

$$E_i = \frac{k_i(\boldsymbol{k} \cdot \boldsymbol{E})}{k^2 - \omega^2 \mu \epsilon_i}. \qquad (6.37)$$

Now, $k_i \times$ (Eq. (6.37)), and using Eq. (6.25), gives

$$k_x E_x = \frac{k_x^2(\boldsymbol{k} \cdot \boldsymbol{E})}{k^2 - \omega^2 \mu \epsilon_x}, \qquad (6.38)$$

$$k_y E_y = \frac{k_y^2(\boldsymbol{k} \cdot \boldsymbol{E})}{k^2 - \omega^2 \mu \epsilon_y}, \qquad (6.39)$$

$$k_z E_z = \frac{k_z^2(\boldsymbol{k} \cdot \boldsymbol{E})}{x^2 - \omega^2 \mu \epsilon_z}. \qquad (6.40)$$

Adding Eqs. (6.38) to (6.40):

$$k_x E_x + k_y E_y + k_z E_z$$

$$= (\boldsymbol{k} \cdot \boldsymbol{E}) \left\{ \frac{k_x^2}{x^2 - \omega^2 \mu \epsilon_x} + \frac{k_y^2}{k^2 - \omega^2 \mu \epsilon_y} + \frac{k_z^2}{k^2 - \omega^2 \mu \epsilon_z} \right\},$$

$$\text{i.e., } \boldsymbol{k} \cdot \boldsymbol{E} = (\boldsymbol{k} \cdot \boldsymbol{E}) \; \{\}$$

or

$$1 = \{ \; \}.$$

Exchanging left and right sides,

$$\frac{k_x^2}{k^2 - \omega^2 \mu \epsilon_x} + \frac{k_y^2}{k^2 - \omega^2 \mu \epsilon_y} + \frac{k_z^2}{k^2 - \omega^2 \mu \epsilon_z} = 1. \qquad (6.41)$$

This is one form of Fresnel's equation of wave normals, representing two sheets of connected surfaces in the 3D k-space. Note that Eq. (6.41) is the surface $w(\boldsymbol{k}) =$ constant. When ω varies, the surface changes. Figure 6.2 shows schematically, at a fixed ω, a section of the surfaces with two symmetrical points of connection. In general, there are two more such points. The origin is the point where the

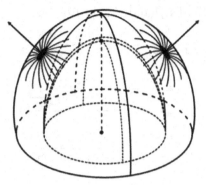

Fig. 6.3. A 3D view of half the surface of the wave normal. The two arrows indicate the optic axes.

material re-radiates. Figure 6.3 shows a 3D version that is cut into two halves. If the k vector (direction of propagation) is along the line joining the origin with a pair of connected points, there is only one solution of k; hence, one phase velocity or one index of refraction. This direction is called the optic axis. There are two such axes in general. Now, if we define a unit vector, S, in the k-space such that its components are the direction cosines of $\hat{k}$,

$$\text{i.e.,} \quad s \equiv \hat{k} = s_x\hat{x} + s_y\hat{y} + s_z\hat{z}, \tag{6.42}$$

where s_x, s_y and s_z are the direction cosines; using

$$k = \frac{2\pi}{\lambda}s = \frac{2\pi}{v/\nu}s = \frac{2\pi\nu}{c/n}s = \frac{\omega}{c}ns, \tag{6.43}$$

we shall transform Eq. (6.41) into another form.
Since

$$k \cdot k = k^2 = \frac{\omega^2}{c^2}n^2 S \cdot S = \frac{\omega^2 n^2}{c^2}, \tag{6.44}$$

$$k_i = \frac{\omega}{c}ns_i(i = x, y, z), \tag{6.45}$$

the first term on the left hand side of Eq. (6.41) becomes

$$\frac{k_x^2}{k^2 - \omega^2 \mu \epsilon_x} = \frac{\frac{\omega}{c} n s_x \cdot \frac{\omega}{c} n s_x}{\frac{\omega^2}{c^2} n^2 - \omega^2 \mu \epsilon_x},$$

$$= \frac{\frac{\omega^2 n^2}{c^2} s_x^2}{\frac{\omega^2}{c^2} n^2 - \frac{\omega^2 \mu \epsilon_0}{\epsilon_0} \epsilon_x},$$

$$= \frac{\omega^2 n^2 s_x^2 / c^2}{\frac{\omega^2}{c^2} n^2 - \frac{\omega^2}{c^2} \frac{\epsilon_x}{\epsilon_0}},$$

$$= \frac{n^2 s_x^2}{n^2 - \frac{\epsilon_x}{\epsilon_0}}. \tag{6.46}$$

Similar expressions can be derived for the other two terms at the left hand side of Eq. (6.41), which now becomes

$$\frac{s_x^2}{n^2 - \epsilon_x / \epsilon_0} + \frac{s_y^2}{n^2 - \epsilon_y / \epsilon_0} + \frac{s_z^2}{n^2 - \epsilon_z / \epsilon_0} = \frac{1}{n^2}. \tag{6.47}$$

This is another form of Fresnel's equation of wave normals showing the explicit dependence on the index n. It describes the same type of surfaces of two connected sheets in the s-(or k-) space, giving two solutions of n in a general direction $\boldsymbol{S}$ (cf. Figs. 6.2 and 6.3 with k changed into $\boldsymbol{S}$), confirming again part of our statement at the beginning of this section.

Method (b)

This method is the direct substitution of Eq. (6.25) into (6.34), writing the latter in matrix form and solving the resulting equation. The latter requires that the determinant of the matrix be zero. Such a condition leads to, after simplification, Eq. (6.41). The calculation is left as an exercise for the reader.

Statement 2

The direction of the electric field vector associated with the k vector is:

$$\begin{pmatrix} \dfrac{k_x}{k^2 - \omega^2 \mu \epsilon_x} \\[2ex] \dfrac{k_y}{k^2 - \omega^2 \mu \epsilon_y} \\[2ex] \dfrac{k_z}{k^2 - \omega^2 \mu \epsilon_z} \end{pmatrix}.$$

Confirmation

Equation (6.37) can be rewritten as:

$$\frac{E_i}{\boldsymbol{k} \cdot \boldsymbol{E}} = \frac{k_i}{k^2 - \omega^2 \mu \epsilon_i} (i = x, y, z). \tag{6.48}$$

Hence,

$$E_x : E_y : E_z = \frac{E_x}{\boldsymbol{k} \cdot \boldsymbol{E}} : \frac{E_y}{\boldsymbol{k} \cdot \boldsymbol{E}} : \frac{E_z}{\boldsymbol{k} \cdot \boldsymbol{E}},$$

$$= \frac{k_x}{k^2 - \omega^2 \mu \epsilon_x} : \frac{k_y}{k^2 - \omega^2 \mu \epsilon_y} : \frac{k_z}{k^2 - \omega^2 \mu \epsilon_z},$$

(using Eq. (6.48)),

i.e., the direction of the vector

$$\boldsymbol{v} \equiv \begin{pmatrix} \dfrac{k_x}{k^2 - \omega^2 \mu \epsilon_x} \\ \dfrac{k_y}{k^2 - \omega^2 \mu \epsilon_y} \\ \dfrac{k_z}{k^2 - \omega^2 \mu \epsilon_z} \end{pmatrix} = \begin{pmatrix} v_x \\ v_y \\ v_z \end{pmatrix} \tag{6.49}$$

is parallel to the direction of the electric field:

$$\boldsymbol{E} = \begin{pmatrix} E_x \\ E_y \\ E_z \end{pmatrix},$$

as shown in Fig. 6.4. Since there are two solutions of k (or n) for a fixed frequency ω and a fixed direction of propagation, $\boldsymbol{v}$ will have two solutions, $\boldsymbol{v}_1$ and $\boldsymbol{v}_2$, and the corresponding electric field $\boldsymbol{E}_1$ and $\boldsymbol{E}_2$ will be parallel to $\boldsymbol{v}_1$ and $\boldsymbol{v}_2$, respectively.

Statement 3

The electric field $\boldsymbol{E}$, the field $\boldsymbol{D}$, and the wave vector $\boldsymbol{k}$ (or $\boldsymbol{S}$) lie in the same plane and $\boldsymbol{D} \perp \hat{k}$.

Confirmation

$\because \nabla \cdot \boldsymbol{D} = 0$, (one of Maxwell's equations),

$\therefore -i\boldsymbol{k} \cdot \boldsymbol{D} = 0$, ($\because \nabla = -i\boldsymbol{k}$ for plane waves, see Eq. (6.30)),

or $\boldsymbol{D} \perp \hat{k}$. $\tag{6.50}$

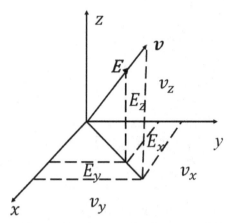

Fig. 6.4.   Vector relationship showing the direction of the electric vector.

This proves the last part of the statement. Because of Eq. (6.34), which we will rewrite as follows:

$$\boldsymbol{k} \times \boldsymbol{k} \times \boldsymbol{E} + \omega^2 \mu \epsilon \boldsymbol{E} = 0, \tag{6.51}$$

$$\hat{k} \times \hat{k} \times \boldsymbol{E} \alpha - \epsilon \boldsymbol{E} = -\boldsymbol{D},$$

$$\therefore -\boldsymbol{D} \parallel \hat{k} \times (\hat{k} \times \boldsymbol{E}). \tag{6.52}$$

Referring to Fig. 6.5(a), it shows pictorially the combination of Eqs. (6.50) and (6.52) in the horizontal plane.

Now,

$$(\hat{k} \times \boldsymbol{E}) \perp \hat{k} \times (\hat{k} \times \boldsymbol{E})$$

by inspection, i.e. $(\hat{k} \times \boldsymbol{E})$ is a vertical vector perpendicular to the horizontal plane, see Fig. 6.5(b). But $\boldsymbol{E} \perp (\hat{k} \times \boldsymbol{E})$ and $\hat{k} \perp (\hat{k} \times \boldsymbol{E})$ by inspection. Hence, $\boldsymbol{E}$ has to be in the same plane as $\hat{k}$, i.e., the horizontal plane (Fig. 6.5(c)). Hence, $\boldsymbol{D}, \boldsymbol{E}, \hat{k}$ lie in the same plane and $\boldsymbol{D} \perp \hat{k}$.

Statement 4

The EM energy propagates in a different (ray) direction than $\hat{k}$, i.e., the Poynting vector $\boldsymbol{S}$ does not coincide with the wave vector $\boldsymbol{k}$, but $\boldsymbol{S}$ is in the same plane as $\boldsymbol{k}$, $\boldsymbol{E}$ and $\boldsymbol{D}$.

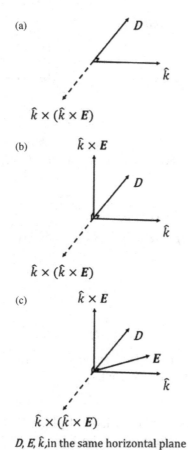

$D, E, \hat{k}$, in the same horizontal plane

Fig. 6.5.   Relationship between $D, \hat{k}, D$ and $\hat{k} \times E$.

Confirmation

We just remind ourselves that $\hat{k}$ is the direction of propagation of the wavefront (surface of constant phase) in the anisotropic medium.

Normally, in an isotropic medium, $S \| k$. However, because E is not parallel to D in an anisotropic medium, S is not parallel to k either, as shown in Fig. 6.6. Figure 6.6(a) shows the result of Statement 3, i.e., they are drawn in the same horizontal plane.

Now,

$$H \perp \hat{k}. \tag{6.53}$$

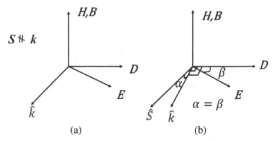

Fig. 6.6.   Relationship between H, B, D, E, $\hat{k}$ and $\hat{S}$.

Since

$$\nabla \cdot B = 0,$$

$$-ik \cdot B = 0, \quad (\text{by Eq. (6.30)}),$$

$$k \perp B$$

or

$$k \perp H'$$

as shown in Fig. 6.6(a) as the vertical vector. Since $S = E \times H$ (definition of the Poynting vector):

$$S \perp H.$$

Hence, S has to be in the horizontal plane (Fig. 6.6(b)). In addition, $S \perp E$ from the definition of the Poynting vector. We can thus see from Fig. 6.6(b) that

$$\alpha = \beta. \tag{6.54}$$

Conclusion

$\hat{S}, \hat{k}, E$ and D are in the same plane and the angle $(\hat{S}, \hat{k})$ is equal to the angle (E, D).

Statement 5

The energy propagation velocity, usually referred to as the group or ray velocity (v_r), makes an angle α with the wavefront propagation velocity or phase velocity (v_p), and

$$\cos \alpha = \frac{v_p}{v_r}. \tag{6.55}$$

Confirmation

We start with Eqs. (6.32) and (6.33), which are rewritten as follows:

$$\boldsymbol{K} \times \boldsymbol{E} = \omega\mu\boldsymbol{H}, \tag{6.56}$$

$$\boldsymbol{K} \times \boldsymbol{H} = -\omega\epsilon\boldsymbol{E} = -\omega\boldsymbol{D}. \tag{6.57}$$

We shall derive a relationship between the total energy density in the EM field and the Poynting vector $\boldsymbol{S}$. This is then used in the definition of $\boldsymbol{v}_{\rm r}$, leading to the desired result. The total energy density in the EM field is the sum of the electric and magnetic energy densities:

$$W = W_e + W_m, \quad (e : \text{electric}, m : \text{magnetic}),$$

$$= \frac{1}{2}\boldsymbol{E} \cdot \boldsymbol{D} + \frac{1}{2}\boldsymbol{B} \cdot \boldsymbol{H},$$

$$= \frac{1}{2}\boldsymbol{E} \cdot \left[-\frac{\boldsymbol{k}}{\omega} \times \boldsymbol{H}\right] + \frac{1}{2}\mu\boldsymbol{H} \cdot \left[\frac{\boldsymbol{k}}{\omega\mu} \times \boldsymbol{E}\right]$$

(using Eqs. (6.56)and (6.57)),

$$= -\frac{1}{2\omega}\boldsymbol{k} \cdot (\boldsymbol{H} \times \boldsymbol{E}) + \frac{1}{2\omega}\boldsymbol{k} \cdot (\boldsymbol{E} \times \boldsymbol{H}).$$

(Where use has been made of the vector identity: $\boldsymbol{a} \cdot (\boldsymbol{b} \times \boldsymbol{c}) = \boldsymbol{b} \cdot (\boldsymbol{c} \times \boldsymbol{a}) = \boldsymbol{c} \cdot (\boldsymbol{a} \times \boldsymbol{b})$.)

Continuing,

$$W = -\frac{1}{2\omega}\boldsymbol{k} \cdot (-\boldsymbol{S}) + \frac{1}{2\omega}\boldsymbol{k} \cdot \boldsymbol{S}, \quad (\because \boldsymbol{S} = \boldsymbol{E} \times \boldsymbol{H}), \tag{6.58'}$$

$$= \frac{\boldsymbol{k} \times \boldsymbol{S}}{\omega}. \tag{6.58}$$

Now, the ray velocity (or group velocity) $\boldsymbol{v}_r$ is defined as the velocity of the transport of the total EM energy:

$$v_{\rm r} = \frac{S}{w} = \frac{\left(\frac{\text{energy}}{\text{sec}}\right) \text{ crossing a unit area } (\perp\boldsymbol{S})}{W\left(\frac{\text{total energy}}{\text{vol.}}\right)}.$$

Hence,

$$v_r = \frac{|\boldsymbol{S}|}{w},$$

$$= \frac{|\boldsymbol{S}|}{\frac{\boldsymbol{k}\cdot\boldsymbol{S}}{\omega}}, \quad \text{(by Eq. (6.58))},$$

$$= \frac{\omega}{\boldsymbol{k}\cdot\hat{\boldsymbol{S}}},$$

$$= \frac{\omega}{K\cos\alpha}, \quad \text{(see Fig. 6.6(b))},$$

$$= \frac{v_p}{\cos\alpha}\cdot\left(\frac{\omega}{k} = \frac{\omega}{\frac{\omega}{c}n} = \frac{c}{n} = v_p\right).$$

Hence,

$$\cos\alpha = \frac{v_p}{v_r}, \quad \text{confirming Eq. (6.55)}.$$

Statement 6

In the $\boldsymbol{S}$-space, there is another surface $\omega(\boldsymbol{S}) = \text{constant}$, giving two group or ray velocities (v_r) or ray indices (n_r). This surface (again of two sheets) is called the ray surface and there is a rule of duality between the ray surface and the surface of wave normals (Statement 1). The origin of the ray surface is the re-radiating point.

Confirmation

We notice that Eq. (6.35), namely:

$$(\boldsymbol{k}\cdot\boldsymbol{E})\boldsymbol{k} - k^2\boldsymbol{E} + \omega^2\mu\epsilon\boldsymbol{E} = 0 \tag{6.59}$$

has led to Fresnel's equation of wave normals. We slightly transform Eq. (6.59) into

$$\boldsymbol{D} = \epsilon\boldsymbol{E} = \frac{k^2}{\omega^2\mu}\{\boldsymbol{E} - \hat{\boldsymbol{K}}(\hat{\boldsymbol{K}}\cdot\boldsymbol{E})\}. \tag{6.60}$$

We shall derive an analogous equation with $\boldsymbol{D}$ and $\boldsymbol{E}$ interchanged so as to obtain the rule of duality. This rule will "logically" give us the ray surface. We define the unit vector:

$$t \equiv \hat{S}. \tag{6.61}$$

Using Eq. (6.60), we have:

$$\boldsymbol{D} \cdot \boldsymbol{t} = \frac{K^2}{\omega^2 \mu} \{ \boldsymbol{E} \cdot \boldsymbol{t} - (\hat{\boldsymbol{k}} \cdot \boldsymbol{t})(\hat{\boldsymbol{k}} \cdot \boldsymbol{E}) \},$$

$$= \frac{-k^2}{\omega^2 \mu} (\hat{\boldsymbol{k}} \cdot \boldsymbol{t})(\hat{\boldsymbol{k}} \cdot \boldsymbol{E}), \quad (\because \hat{\boldsymbol{k}} \perp \boldsymbol{t}). \tag{6.62}$$

Using again Eqs. (6.60) and (6.62),

$$\boldsymbol{D} - (\boldsymbol{D} \cdot \boldsymbol{t})\boldsymbol{t} = \frac{k^2}{\omega^2 \mu} \{ \boldsymbol{E} - \hat{\boldsymbol{k}}(\hat{\boldsymbol{k}} \cdot \boldsymbol{t}) + (\hat{\boldsymbol{k}} \cdot \boldsymbol{t})(\hat{\boldsymbol{k}} \cdot \boldsymbol{E})\boldsymbol{t} \},$$

$$= \frac{k^2}{\omega^2 \mu} \{ (E\cos\alpha)\hat{\boldsymbol{D}} + (E\sin\alpha)\hat{\boldsymbol{k}} - (E\sin\alpha)\hat{\boldsymbol{k}} + (\cos\alpha)$$

$$(E\sin\alpha)\boldsymbol{t} \}, \text{(Fig. 6.6(b) with } \alpha = \beta),$$

$$= \frac{k^2}{\omega^2 \mu} \{ (E\cos\alpha)(\cos\alpha)\hat{\boldsymbol{E}} + (E\cos\alpha)(\sin\alpha)(-\boldsymbol{t})$$

$$+(E\cos\alpha\sin\alpha)\boldsymbol{t} \}, \text{(Fig. 6.6(b) with } \alpha = \beta),$$

$$= \frac{k^2}{\omega^2 \mu} \boldsymbol{E}\cos^2\alpha. \tag{6.63}$$

But,

$$\cos^2 \alpha = \left(\frac{v_{\mathrm{p}}}{v_{\mathrm{r}}} \right)^2 = \frac{(c/n)^2}{(c/n_{\mathrm{r}})^2} = \frac{n_{\mathrm{r}}^2}{n^2}, \tag{6.64}$$

where n_r is the ray index:

$$v_r = \frac{c}{n_r}.$$

Hence, Eq. (6.63) becomes:

$$\boldsymbol{D} - (\boldsymbol{D} \cdot \boldsymbol{t})\boldsymbol{t} = \frac{k^2}{\omega^2 \mu} \boldsymbol{E} \left(\frac{n_r^2}{n^2} \right),$$

$$= \frac{\omega^2 n^2/c^2}{\omega^2 \mu} \boldsymbol{E} \left(\frac{n_r^2}{n^2} \right),$$

$$\text{i.e., } \boldsymbol{E} = \frac{\mu c^2}{n_r^2} \boldsymbol{D} - (\boldsymbol{D} \cdot \boldsymbol{t})\boldsymbol{t}. \tag{6.65}$$

Comparing this equation with Eq. (6.60), which is rewritten in the following form:

$$D = \frac{n^2}{c^2\mu}\{E - (E \cdot \hat{k})\hat{k}\}. \tag{6.66}$$

We can write down the following rule of duality to transform from the k–space to the S– (or t–) space:

$$k - space \rightarrow S(or\ t)space,$$

$$D \rightarrow E,$$

$$\frac{1}{\mu} \rightarrow \mu,$$

$$\frac{1}{c} \rightarrow c,$$

$$n \rightarrow \frac{1}{n_r},$$

$$\hat{k} \rightarrow -t,$$

$$\varepsilon_i \rightarrow \frac{1}{\varepsilon_i}(i = x, y, z).$$

Since Eq. (6.66) has led to Fresnel's equation of wave normal, the ray surface can be obtained by simply applying the above transformation rule to the equation of the wave normal. For example, Eq. (6.47).

$$\frac{S_x^2}{n^2 - \epsilon_x/\epsilon_0} + \frac{S_y^2}{n^2 - \epsilon_y/\epsilon_0} + \frac{S_z^2}{n^2 - \epsilon_z/\epsilon_0} = \frac{1}{n^2} \tag{6.67}$$

becomes

$$\frac{t_x^2}{\frac{1}{n_r^2} - \frac{1}{\epsilon_x \epsilon_0}} + \frac{t_y^2}{\frac{1}{n_r^2} - \frac{1}{\epsilon_y \epsilon_0}} + \frac{t_z^2}{\frac{1}{n_r^2} - \frac{1}{\epsilon_z \epsilon_0}} = n_r^2. \tag{6.68}$$

Equation (6.68) is the equation of the ray surface. It again is a surface of two shells similar to Figs. (6.2) and (6.3), with k (or S) replaced by t. Thus, for a general direction t emanating from the origin where the material reradiates, there are two solutions for n_r. (Note: There is a relationship between the ray surface and the surface of the wave normals, namely, that the latter is the pedal surface of the former. The reader is referred to Born and Wolf.)

Statement 7

Using the index ellipsoid and the inverse (or Fresnel) ellipsoid, one can show that $\boldsymbol{D}_1 \perp \boldsymbol{D}_2$ and $\boldsymbol{E}_1 \perp \boldsymbol{E}_2$, respectively, where 1 and 2 refer to the two solutions of the EM wave propagation in the anisotropic medium (governed by the surface of wave normals and the ray surface).

Confirmation

We shall first define the index ellipsoid and the inverse (or Fresnel) ellipsoid. The electric part of the energy density of the EM wave is:

$$W_e = \frac{1}{2} \boldsymbol{E} \cdot \boldsymbol{D}$$

$$2W_e = (E_x, E_y, E_z) \begin{pmatrix} \epsilon_x & 0 & 0 \\ 0 & \epsilon_y & 0 \\ 0 & 0 & \epsilon_z \end{pmatrix} \begin{pmatrix} E_x \\ E_y \\ E_z \end{pmatrix},$$

$$= \epsilon_x E_x^2 + \epsilon_x E_y^2 + \epsilon_z E_z^2, \text{ (inverse or Fresnel ellipsoid)}, \quad (6.69)$$

$$= \frac{D_x^2}{\epsilon_x} + \frac{D_y^2}{\epsilon_y} + \frac{D_z^2}{\epsilon_z}, \text{ (index ellipsoid)}. \quad (6.70)$$

Note that the Fresnel ellipsoid represents the surface of $W_e(\mathbf{E}) = $ constant, whereas the index ellipsoid represents the surface of $W_e(\mathbf{D}) = $ constant.

(a) Index ellipsoid or optical indicatrix

Using the definitions:

$$\boldsymbol{r} \equiv (x, y, z),$$

$$\equiv \frac{\boldsymbol{D}}{\sqrt{2W_e \epsilon_0}}, \text{ (dimensionless)}, \quad (6.71)$$

$$n_i^2 = \frac{\epsilon_i}{\epsilon_0}, \quad (i = x, y, z). \quad (6.72)$$

Equation (6.70) becomes:

$$\frac{x^2}{n_x} + \frac{y^2}{n_y} + \frac{z^2}{n_z} = 1, \quad (6.73)$$

which is the index ellipsoid with major axes $2n_x$, $2n_y$ and $2n_z$ along

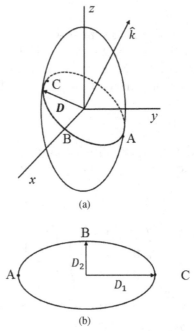

(a)

(b)

Fig. 6.7.   Index ellipsoid.

the x, y and z axes. In general, for a constant W_e, any point on the surface of the index ellipsoid is a solution of the index n or the field D because x, y and z are proportional to D_x, D_y and D_z (Eq. (6.71)). However, as shown in Fig. 6.7(a) for a given direction of propagation k of the wavefront, the field D associated with the wave vector k should lie in the plane perpendicular to k, because $D \perp k$ (Eq. (6.50)). This plane intersects the surface of the index ellipsoid in an ellipse A, B and C. This ellipse is redrawn in Fig. 6.7(b). Any vector drawn from the origin of the ellipsoid to the ellipse is a solution of D in the material in the infinitesimal region surrounding the origin. Because D is the response of the material to the traveling plane EM wave, it is also a traveling wave of the same frequency (i.e., linear response), and the above solution means that the tip of the vector D traces out the ellipse; in other words, D is elliptically polarized. As will be seen in Chapter 7, any elliptically polarized vector can be described by the vector combination of two orthogonal vector waves with two wave vectors k_1 and k_2, both propagating in the same direction of

$\hat{k}$ (or $\hat{S}$), which we denote by the z'-direction here:

$$D_1 = D_{10} \cos(\omega t - k_1 z'),$$
$$D_2 = D_{20} \cos(\omega t - k_2 z'),$$

where z' is the position of the point of response (origin of the index ellipsoid) with respect to an arbitrary coordinate system. The above two equations can be rewritten in the more familiar form (used in Chapter 7):

$$D_1 = D_{10} \cos(\omega t - k_1 z'),$$
$$D_2 = D_{20} \cos(\omega t - k_1 z' + \epsilon),$$

where

$$\epsilon \equiv \text{relative phase difference between } D_1 \text{ and } D_2,$$
$$\equiv (k_1 - k_2)z'.$$

$2D_{10}$ and $2D_{20}$ are the principal axes of the ellipse. Mathematically, it can indeed be rigorously proved that the lengths of the principal axes of the ellipse are the two solutions of the index n (or v, k or D) given by the surface of wave normals, for a propagation direction $\hat{k}$. The fields D corresponding to these two solutions are along the principal axes, and thus, are orthogonal (see Born and Wolf for details). We use the convention of the right-hand rotation so that $(D_1 \times D_2)$ points in the direction of k. These two waves D_1 and D_2 are sometimes called the normal modes in the medium for a particular propagation direction $\hat{k}$. In all, one can thus say that for a given direction $\hat{k}$, there are two solutions of D, namely, D_1 and D_2, and $D_1 \perp D_2$. Each of these fields (D_1 and D_2) propagates with a different velocity that is governed by the surface of wave normals.

(b) Fresnel or inverse ellipsoid

Equation (6.69) can be rewritten as:

$$\frac{E_x^2}{\left(\sqrt{\frac{2W_e}{\epsilon_x}}\right)^2} + \frac{E_y^2}{\left(\sqrt{\frac{2W_e}{\epsilon_y}}\right)^2} + \frac{E_z^2}{\left[\sqrt{\frac{2W_e}{\epsilon_z}}\right]^2} = 1. \tag{6.74}$$

Equation (6.74) is an ellipsoid whose surface points give solutions to the electric field E. However, as in the case of the index ellipsoid, for

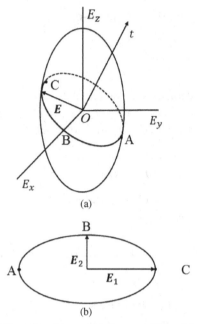

(a)

(b)

Fig. 6.8.   Fresnel or inverse ellipsoid.

a given direction t (or S) (see Fig. 6.8(a)), the solution of E should lie in the plane perpendicular to t, since $E \perp \hat{S}$ (or t) by the definition of the Poynting vector. The solution becomes a vector from 0 terminating at the intersection ellipse, which is redrawn in Fig. 6.8(b). Similar to the case of D, this electric field can be "decomposed" into two orthogonal fields E_1 and E_2 so that $E_1 \times E_2$ is in the direction of t (using the right-hand convention). We again conclude that for a given direction t, there are two solutions of E, namely E_1 and E_2, and $E_1 \perp E_2$. Each of these fields (E_1 and E_2) propagates with a different velocity governed by the ray surface.

Statement 8

Given a direction $\hat{k}$ of the wave vector, the fields associated with k are such that (D_1, E_1) and (D_2, E_2) lie in two perpendicular planes with k as the intersection line between the planes.

Confirmation

This is easily seen in Fig. 6.9(a). According to Statement 3, D, E and k lie in the same plane. Hence. D_1, E_1 and k lie in one plane

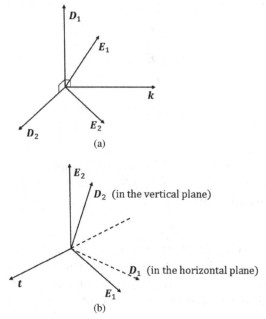

Fig. 6.9. Relationship between $\boldsymbol{E}_1, \boldsymbol{E}_2, \boldsymbol{D}_1, \boldsymbol{D}_2, \hat{\boldsymbol{k}}$ and $\boldsymbol{t}$.

and $\boldsymbol{D}_2$, $\boldsymbol{E}_2$ and $\boldsymbol{k}$ lie in another plane so that $\boldsymbol{k}$ is the intersection line of the two planes.

Since $\boldsymbol{D}_1 \perp \boldsymbol{D}_2$ (Statement 7, first part), the two planes must be perpendicular to each other.

Statement 9

Given a direction $\hat{\boldsymbol{S}}$ (or $\boldsymbol{t}$) of the Poynting vector, the fields associated with $\boldsymbol{S}$ are such that $(\boldsymbol{E}_1, \boldsymbol{D}_1)$ and $(\boldsymbol{E}_2, \boldsymbol{D}_2)$ lie in two perpendicular planes with $\boldsymbol{t}$ (or $\hat{\boldsymbol{S}}$) as the intersection line between the two planes.

Confirmation

As shown in Fig. 6.9(b), since $\boldsymbol{t}, \boldsymbol{D}, \boldsymbol{E}$ and $\boldsymbol{k}$ in general lie in the same plane (Statement 4), we should have $\boldsymbol{t}, \boldsymbol{E}_1$ and $\boldsymbol{D}_1$ lying in one plane and $\boldsymbol{t}$, $\boldsymbol{E}_2$ and $\boldsymbol{D}_2$ lying in another. Thus, $\boldsymbol{t}$ forms the intersection line between the two planes. Since $\boldsymbol{E}_1 \perp \boldsymbol{E}_2$ (Statement 7), the two planes should be perpendicular.

Statement 10

$$D_1 \perp E_2,$$
$$D_2 \perp E_1,$$
$$\hat{k} \cdot (E_1 \times H_2) = 0,$$
$$\hat{k} \cdot (E_2 \times H_1) = 0.$$

Confirmation

This is left as an <u>exercise</u> for the reader. He merely has to draw the appropriate H vectors onto Fig. 6.9(a) and the resultant figure will prove the above relationship.

Statement 11

Given a wave vector $\hat{k}$, there are in general two possible t's lying in two planes perpendicular to each other with $\hat{k}$ as the intersection line of the two planes. Each of the two Poynting vectors S_1 and S_2 (corresponding to the two t's) terminates on a sheet of the ray surface.

Confirmation

According to Statement 8, given a wave vector k, there are two pairs of (E_1, D_1) and (E_2, D_2) lying in two perpendicular planes with k as the intersection line. Since t, D, E and k all lie in the same plane (Statement 4), (t_1, E_1, D_1) and (t_2, E_2, D_2) should lie in two perpendicular planes with k as the intersection line (Fig. 6.10(a)).

Statement 12

Given a Poynting vector's direction t, there are in general two possible k's lying in two planes perpendicular to each other with t as the intersection line of the two planes. Each of the two k's terminates on one sheet of the surface of wave normal.

Confirmation

As in Statement 11, using Statements 9 and 4, the above statement can be seen easily (Fig. 6.10(b)).

Summary

Because of the anisotropy, the dielectric constant ϵ becomes a tensor (matrix), which gives different values of ϵ in different directions.

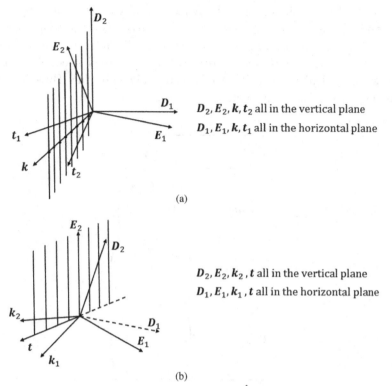

D_2, E_2, k, t_2 all in the vertical plane

D_1, E_1, k, t_1 all in the horizontal plane

(a)

D_2, E_2, k_2, t all in the vertical plane

D_1, E_1, k_1, t all in the horizontal plane

(b)

Fig. 6.10. Possible relationship between D, E, $\hat{k}$ and t. The plane containing k_1 and k_2 is perpendicular to the plane containing D_1 and D_2. In (a), E_1 is not perpendicular to E_2, whereas in (b), D_1 is not perpendicular to D_2. Note that the origin is the source of radiation.

The resultant fields D and E generated by the interaction of a plane EM wave with the medium are not parallel. Maxwell's equations lead to the conclusion that when a plane EM wave propagates in an anisotropic medium, it decomposes into two plane waves of different propagation velocities, each of them linearly polarized, and the fields perpendicular to each other. ($D_1 \perp D_2$ in the direction of k and/or $E_1 \perp E_2$ in the direction of t or $\widehat{S}$.)

6.3 Classification of Anisotropic Material Optically

Most anisotropic materials are crystals. It is the different types of crystal symmetries that give rise to anisotropy. A full understanding of the matter requires a good knowledge of crystallography; we cannot go into such details because it would be outside the scope of this book. Rather, we assume that all crystals of interest have already been studied and classified according to the crystal specialists. The following optical classification of anisotropic materials thus ignores the crystal symmetry aspect.

Classification	ϵ_{ij}
Isotropic	$\begin{pmatrix} \epsilon & 0 & 0 \\ 0 & \epsilon & 0 \\ 0 & 0 & \epsilon \end{pmatrix}$
Uniaxial	$\begin{pmatrix} \epsilon_x & 0 & 0 \\ 0 & \epsilon_x & 0 \\ 0 & 0 & \epsilon_z \end{pmatrix}$
Biaxial	$\begin{pmatrix} \epsilon_x & 0 & 0 \\ 0 & \epsilon_y & 0 \\ 0 & 0 & \epsilon_z \end{pmatrix}$

(1) Isotropic material

   This is the normal material (dielectrics) one assumes in any book on electromagnetism. Its interaction with the EM wave is well known.

(2) Uniaxial material

   According to our definition:

$$\epsilon_y = \epsilon_x. \tag{6.75}$$

We define:

$$n_o^2 \equiv \epsilon_x/\epsilon_0, \tag{6.76}$$

$$n_e^2 \equiv \epsilon_z/\epsilon_0. \tag{6.77}$$

Substituting Eqs. (6.75) to (6.77) into the equation of wave normals (Eq. (6.41)), and using

$$k^2 = k_x^2 + k_y^2 + k_z^2, \tag{6.78}$$

$$= \frac{\omega^2}{c^2} n^2, \tag{6.79}$$

one obtains (the detailed calculation is left as an <u>exercise</u>):

$$(n^2 - n_o^2) \left\{ (k_x^2 + k_y^2)n_o^2 n^2 + k_z^2 n_e^2 n^2 - n_o^2 n_e^2 \frac{\omega^2}{c^2} n^2 \right\} = 0. \tag{6.80}$$

This leads to two simultaneous solutions:

$$n^2 - n_o^2 = 0 \tag{6.81}$$

and

$$\frac{k_x^2 + k_y^2}{n_e^2} + \frac{k_z^2}{n_o^2} = \frac{\omega^2}{c^2}. \tag{6.82}$$

Equation (6.81) can be transformed into:

$$k_x^2 + k_y^2 x + k_z^2 \equiv k^2 = \frac{\omega^2 n_o^2}{c^2}, \quad (\text{using Eq. (6.79)})$$

or

$$\frac{k_x^2}{\frac{x_0}{\frac{\omega^2 n_o^2}{c^2}}} + \frac{x_y^2}{\frac{w^2 n_o^2}{c^2}} + \frac{x_z^2}{\frac{w^2 n_o^2}{c^2}} = 1. \tag{6.83}$$

And Eq. (6.82) is rewritten:

$$\frac{k_x^2}{\frac{\omega^2 n_e^2}{c^2}} + \frac{k_y^2}{\frac{\omega^2 n_e^2}{c^2}} + \frac{k_z^2}{\frac{\omega^2 n_o^2}{c^2}} = 1. \tag{6.83'}$$

The simultaneous solutions in the $k - space$ (wave normals) are thus a spherical surface (Eq. (6.83)) and an ellipsoid (Eq. (6.83')) whose axis of rotational symmetry is around the z-axis. Note that the length of the principal axis of the ellipsoid along the z-direction is equal to the radius of the sphere ($\omega n_o/c$). Thus, the ellipsoid and the sphere should touch. Depending on whether n_e is larger or smaller than n_0, we have the surface of the wave normals shown schematically in

Fig. 6.11. We define the anisotropy as positive (negative) uniaxial if $n_e > (<)n_o$. n_o and n_e are defined as the indices of the ordinary (also known as "o") and extraordinary (also known as "e") waves. (We notice again that the origin is the re-radiating point.)

The polarizations of the ordinary and extraordinary waves can be investigated by using the index ellipsoid. Because it is uniaxial, we define that the indices in the x- and y-directions are equal to n_o. We thus have the ellipsoid with the z-axis as the rotational symmetric axis shown in Fig. 6.12. Its equation is:

$$\frac{D_x^2}{2\epsilon_x w_e} + \frac{D_y^2}{2\epsilon_x w_e} + \frac{D_z^2}{2\epsilon_z w_e} = 1, \quad \text{(from Eq. (6.70))}. \tag{6.84}$$

Because of the rotational symmetry, any propagation direction $\hat{k}$ of the wave vector is symmetric around z; i.e., for an arbitrary $\hat{k}$, one can choose the y–z plane to coincide with $\hat{k}$ without loss of generality. The y–z plane (containing $\hat{k}$ and the z-axis of symmetry) is

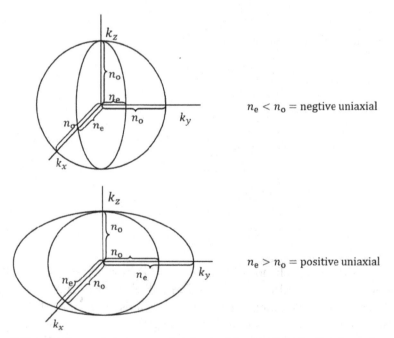

Fig. 6.11.   Surfaces of wave normals of uniaxial materials. In the above figures, to be precise, each n should be multiplied by ω/c. We have neglected this for the sake of clarity.

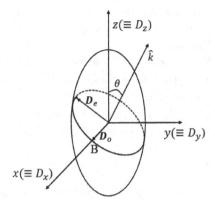

Fig. 6.12.   Index ellipsoid of a uniaxial material.

called the principal plane. Any change in direction of $\hat{k}$ becomes the change in the angle θ in the y–z plane. The ellipse formed by the intersection of the plane perpendicular to $\hat{k}$, and the index ellipsoid defines the field $\boldsymbol{D}$, having two components along the principal axes of the ellipse. Because this ellipse is symmetrical about the principal plane (y–z plane), one of the principal axes of the ellipse is always perpendicular to the principal plane and the other lies in the plane. We denote them by $\boldsymbol{D}_\mathrm{o}$ and $\boldsymbol{D}_\mathrm{e}$, respectively. Whatever the direction of $\hat{k}$ is (i.e., changing θ), $\boldsymbol{D}_\mathrm{o}$, which is now along the x-axis in Fig. 6.12, is unchanged; hence, it is the isotropic ordinary wave. $\boldsymbol{D}_\mathrm{e}$ is perpendicular to $\boldsymbol{D}_\mathrm{o}$. Since $\boldsymbol{D}_\mathrm{o}$ is isotropic:

$$\boldsymbol{D}_\mathrm{o} \parallel \boldsymbol{E}_\mathrm{o},$$

i.e., $\boldsymbol{E}_\mathrm{o}$ is in the x-direction. By Statement 3 of the previous section, $\hat{k}$, $\boldsymbol{D}_\mathrm{e}$ and $\boldsymbol{E}_\mathrm{e}$ should lie in the same $y - z$ plane. Hence, $\boldsymbol{E}_\mathrm{o}\perp\boldsymbol{E}_\mathrm{e}$. That is to say, when an EM wave interacts with a uniaxial medium, it generates two waves with two different propagation velocities. Both the ordinary (isotropic) wave and the extraordinary waves are linearly polarized and the two polarizations $\boldsymbol{E}_o$ and $\boldsymbol{E}_e$ are orthogonal.

At $\theta = 0$, ($\hat{k}$ coincides with the z-axis), $|\boldsymbol{D}_e| = |\boldsymbol{D}_0|$, because the ellipse becomes a circle, i.e., when the wave vector is in the z-direction, there is only one wave velocity. This means that the z-axis is the optic axis. It also means that $\boldsymbol{E}_o$, the polarization of the ordinary wave, is always perpendicular to the optic axis. The electric vector $\boldsymbol{E}_e$ lies in the y–z plane, i.e., in the principal plane containing $\boldsymbol{k}$ and the optic axis.

Because $\boldsymbol{D}_o \parallel \boldsymbol{E}_o$ in the ordinary wave, the angle between $\boldsymbol{D}_o$ and $\boldsymbol{E}_o$ is zero. This means that $\boldsymbol{k}_o \parallel \boldsymbol{S}_o$ (Statements 4 and 5). However, the extraordinary wave will always have $\boldsymbol{D}_e \nparallel \boldsymbol{E}_e$, and hence $\boldsymbol{k}_e \nparallel \boldsymbol{S}_e$, i.e., the wave vector $\boldsymbol{k}_e$ is in a different direction than the Poynting vector $\boldsymbol{S}_e$. (These can also be analyzed using Statement 2.)

(3) Biaxial material

This is the general case for optical anisotropy. We assume:

$$\epsilon_x < \epsilon_y < \epsilon_z \tag{6.85}$$

or

$$V_x > V_y > V_z \quad \left(\text{since } \frac{\epsilon_x}{\epsilon_0} \equiv n_x^2, \text{ and } V_x = \frac{c}{n_x}\right). \tag{6.85'}$$

In Statement 1, we mentioned a second method of deriving Fresnel's equation of wave normals, and have left it as an <u>exercise</u> for the readers. Those who have done this will first arrive at the following equation:

$$\begin{vmatrix} \omega^2 \mu \epsilon_x - k_y^2 - k_z^2 & k_x k_y & k_x k_z \\ k_y k_x & \omega^2 \mu \epsilon_y - k_x^2 - k_z^2 & k_y k_z \\ k_z k_x & k_z k_y & \omega^2 \mu \epsilon_z - k_x^2 - k_y^2 \end{vmatrix} = 0. \tag{6.86}$$

Equation (6.86) is essentially an equation of the wave normals in a "confused" form. We could examine it more closely by looking at the projections of the equation on the $k_x = 0$, $k_y = 0$, and $k_z = 0$ planes successively (i.e., on the planes perpendicular to the principal dielectric axes; see Section 6.2, Statement 1). We again leave it as an <u>exercise</u> to the reader to simplify Eq. (6.86) under the above three separate conditions, thus obtaining the projections on the $k_x = 0$, $k_y = 0$ and $k_z = 0$ planes as shown in Fig. 6.13(a), 6.13(b) and 6.13(c), respectively. In (a), a circle lies inside an oval; in (b), a circle intersects an oval; and in (c), an oval is inside a circle. We consider

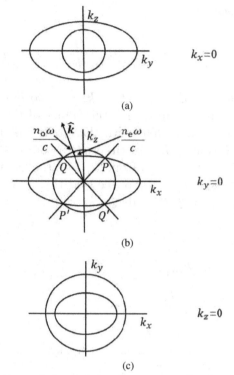

$k_x = 0$

(a)

$k_y = 0$

(b)

$k_z = 0$

(c)

Fig. 6.13. Projections of the surface of wave normals of a biaxial material on the $x - z$, $y - z$ and $x - z$ planes.

only the case in Fig. 6.13(b). Simplification of Eq. (6.86) gives two solutions simultaneously:

$$\frac{k_x^2}{n_y^2} + \frac{k_z^2}{n_y^2} = \frac{\omega^2}{c^2} \tag{6.87}$$

and

$$\frac{k_x^2}{n_z^2} + \frac{k_z^2}{n_x^2} = \frac{\omega^2}{c^2}. \tag{6.88}$$

Equation (6.87) and (6.88) are represented in Fig. 6.13(b) by the circle and the ellipse, respectively. (Note: An ellipse is a special case of an oval.) The two intersect because of our assumption that e $\epsilon_x <$

$\epsilon_y < \epsilon_z$, which means:

$$n_x < n_y < n_z, \quad (\because n_i^2 = \frac{\epsilon_i}{\epsilon_0}, i = x, y, z).$$

In a general direction of propagation $\hat{\boldsymbol{k}}$, the two intersection points with the ellipse and the circle give the values of the two wave vectors. For the intersection point with the circle, the direction of the polarization (direction of $\boldsymbol{E}_o$) of the wave associated with the wave vector $(n_o\omega/c)$ can be obtained from Eq. (6.49) of Statement 2 of the previous section, using $k_y = 0$. We see that the x- and z-components have finite values under the condition $k_y = 0$ while the y-component becomes:

$$\frac{k_y}{k^2 - \omega^2 \mu \epsilon_y} = \frac{k_y}{k^2 - \omega^2 \mu \epsilon_0 \frac{\epsilon_y}{\epsilon_0}}$$

$$= \frac{k_y}{k_x^2 + k_z^2 + k_y^2 - \frac{\omega^2}{c^2}n_y^2}$$

$$= \frac{k_y}{\frac{\omega^2}{c^2}n_y^2 + k_y^2 - \frac{\omega^2}{c^2}n_y^2} \quad \text{(using Eq. (6.87) for the circle)}$$

$$= \frac{k_y}{k_y^2}$$

$$= \frac{1}{k_y} \to \infty \text{ as } k_y \to 0$$

Hence, the direction of $\boldsymbol{E}_o$ (or polarization) becomes:

$$\begin{pmatrix} \text{finite} \\ \infty \\ \text{finite} \end{pmatrix},$$

i.e., the polarization for the isotropic wave $\boldsymbol{E}_o$ is in the y-direction, or perpendicular to the $x - z$ plane. This also means that $\boldsymbol{E}_o$ is always perpendicular to the optic axes POP' and QOQ'. For the extraordinary wave, i.e., the intersection points of $\hat{\boldsymbol{k}}$ and the ellipse (Eq. (6.88)), the same consideration leads to the following direction of $\boldsymbol{E}_e$ (using Eqs. (6.88) and (6.49)):

$$\begin{pmatrix} \text{finite} \\ 0 \\ \text{finite} \end{pmatrix}.$$

Hence, the polarization $\boldsymbol{E}_e$ of the extraordinary wave lies in the $x - z$ plane. Since $\boldsymbol{E}_o$ is perpendicular to the $x - z$ plane:

$$\boldsymbol{E}_o \perp \boldsymbol{E}_e. \tag{6.89}$$

Similarly, one can show that for propagation directions in the $k_x = 0$ and $k_z = 0$ planes, Eq. (6.89) is always valid. (This is again left as an <u>exercise</u>.) We thus conclude that in a general direction of propagation $\boldsymbol{k}$ in a biaxial material, there are two waves that propagate with two different velocities. Their polarizations are always linear and orthogonal. This conclusion is valid in any general direction as demonstrated by Fig. 6.8, using the Fresnel ellipsoid.

6.4 Double Refraction at a Boundary

When a plane wave of wave vector $\boldsymbol{k}_i$ propagates from an isotropic medium into an anisotropic medium, it will generate two waves with different velocities of propagation (or different wave vectors) in two different directions. The reason why this happens can be seen from the following analysis.

Consider point 0 in Fig. 6.14(a), where the incident wave vector $\boldsymbol{k}_i$ intersects the boundary separating the isotropic and anisotropic media. When interacting with the incident EM wave, the material at the anisotropic side of the boundary will re-radiate according to the "rules" (or statements) discussed in Section 6.2. That is to say, the re-radiated wave vectors $\boldsymbol{k}'$ will terminate on the surface of wave normals governed by Eq. (6.41), which is rewritten as follows:

$$\frac{k_x^2}{k'^2 - \omega^2 \mu_x} + \frac{k_y^2}{k'^2 - \omega^2 \mu_y} + \frac{k_z^2}{k'^2 - \omega^2 \mu_z} = 1.$$

We use this particular form of wave normals because we shall be dealing with wave vectors rather than refractive indices or velocities. Figure 6.14(a) shows half of the surface of wave normals inside the

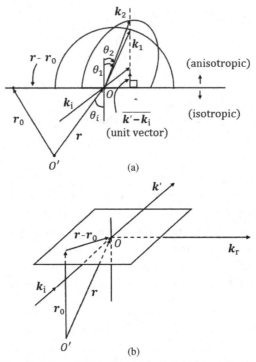

Fig. 6.14. Double refraction. Point O, which should be at the common center of the half-ellipse and half-circle, is purposely displaced, leaving more space for the drawing.

anisotropic medium. The surface of wave normals is of two sheets, of course.

Now, referring to Fig. 6.14(b), from an arbitrary reference frame $0'$, the point 0 is at the position r. The boundary conditions at 0 require that the tangential components of E be continuous across the boundary,

i.e., $(E_i)_t \exp[j(\omega t - k_i \cdot r)] + (E_r)_t \exp[j(\omega t - k_r \cdot r + \epsilon_r)]$

$$= (E')_t \exp[j(\omega t - k' \cdot r + \epsilon')], \qquad (6.90)$$

where the subscript "t" means "tangential", "i" and "r" mean "incident" and "reflected", "$\bullet$" (prime) means transmitted $j = \sqrt{-1}$, and ϵ is the relative phase difference of the reflected (ϵ_r) or transmitted (ϵ') wave with respect to the incident wave. For a given k_i, ϵ_r and ϵ' are constants. Equation (6.90) is valid only if all the phases are equal. In particular,

$$\text{i.e.,} \quad \omega t - \boldsymbol{k}_i \cdot \boldsymbol{r} = \omega t - \boldsymbol{k}' \cdot \boldsymbol{r} + \epsilon', \tag{6.91}$$

$$\boldsymbol{r} \cdot (\boldsymbol{k}_i - \boldsymbol{k}') = \epsilon' = \text{constant} \equiv \boldsymbol{r}_0 \cdot (\boldsymbol{k}_i - \boldsymbol{k}'), \tag{6.92}$$

where $\boldsymbol{r}_0$ is a constant vector terminating on the boundary plane at an arbitrary point while $(\boldsymbol{k}_i - \boldsymbol{k}')$ is a constant vector for a given $\boldsymbol{k}_i$. Equation (6.92), which is the equation of a plane, becomes:

$$(\boldsymbol{r} - \boldsymbol{r}_0) \cdot (\boldsymbol{k}' - \boldsymbol{k}_i) = 0. \tag{6.93}$$

Since $\boldsymbol{r}$ and $\boldsymbol{r}_0$ both terminate on the boundary plane, and $\boldsymbol{r}$ is a variable vector (varies as the position 0 varies), $(\boldsymbol{r} - \boldsymbol{r}_0)$ is a general vector in this plane. Equation (6.93) means that the vector $(\boldsymbol{k}' - \boldsymbol{k}_i)$ is perpendicular to the boundary plane spanned by $(\boldsymbol{r} - \boldsymbol{r}_0)$. But the vector $(\boldsymbol{k}' - \boldsymbol{k}_i)$ intersects the surface of wave normals at two points (Fig. 6.14(a)) because $\boldsymbol{k}'$, the refracted wave vector, should terminate at the surface of wave normals. Hence, there will be two $\boldsymbol{k}'$s or transmitted wave vectors, which we denote by $\boldsymbol{k}_1$, $\boldsymbol{k}_2$.

Again, according to Eq. (6.92), we should have (Fig. 6.14(a)):

$$\boldsymbol{k}_i \sin \theta_i = k' \sin \theta' \tag{6.94}$$

where

$$\theta_i = \begin{pmatrix} \theta_1 \\ \theta_2 \end{pmatrix} \tag{6.95}$$

and Θ_i is the angle of incidence while Θ_1 and Θ_2 are the two angles of refraction.

Rewriting Eq. (6.94):

$$\boldsymbol{k}_i \sin \theta_i = \begin{pmatrix} k_1 \sin \theta_1 \\ k_2 \sin \theta_2 \end{pmatrix}, \tag{6.96}$$

$$n_i \sin \theta_i = \begin{pmatrix} n_1 \sin \theta_1 \\ n_2 \sin \theta_2 \end{pmatrix} \text{ since } k = \omega n / c. \tag{6.96'}$$

Equations (6.96) and (6.96′) are Snell's law for refraction at an anisotropic boundary. It should be noted that $|\boldsymbol{k}_2|$ or n_2 depends on θ_2 while $|\boldsymbol{k}_1|$ or n_1, being isotropic, is independent of θ_1. We stress that $\hat{\boldsymbol{k}}_1$ and $\hat{\boldsymbol{k}}_2$ denote the propagation direction of two wavefronts.

The ray or the Poynting vector, namely the EM energy transport directions, are those associated with $\hat{k}_1$ and $\hat{k}_2$ according to Statements 4, 5 and 6 of Section 6.2. We denote the two ray directions as t_1 and t_2, each having a fixed angular relationship with $\hat{k}_1$ and $\hat{k}_2$, respectively, according to Statement 5. Then, if k_i represents the direction of an incident laser beam, the two refracted beams that will be observed (i.e., energy transport) will be along t_1 and t_2, not $\hat{k}_1$ and $\hat{k}_2$. In the drawing in Fig. 6.14(a), since k_1 terminates on the spherical surface, it represents the isotropic wave. Hence, $k_1 \parallel t_1$.

One can also show that k_1, k_2, k_i and k_r all lie in the same plane of incidence using the argument that is usually used in analyzing refraction across the interface of two isotropic media. We leave this as an <u>exercise</u> to the reader.

Finally, according to the discussion in Section 6.3, the electric fields E_1, E_2 associated with k_1, k_2 are perpendicular because E_1 belongs to the ordinary (isotropic) wave while E_2 belongs to the extraordinary wave.

6.5   Conical Emission From a Biaxial Crystal

When a wave propagates along the optic axis of a biaxial crystal, the wave degenerates into a cone of rays. Figure 6.15 shows the

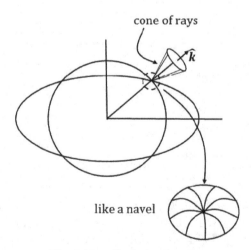

Fig. 6.15.   Conical refraction.

same cross-section as Fig. 6.13(b). When the wave vector k is in the direction of the optic axis, the ray or Poynting vector will be in the direction of v_r or group velocity.

Since

$$v_r = \nabla_k \omega \left(see\, chap.\, 1, v = \frac{d\omega}{dk} \right) \tag{6.97}$$

and the wave surface is in fact

$$\omega(k) = \text{constant}, \tag{6.98}$$

v_r is normal to the surface of wave normals. In particular, in the vicinity of the optic axis, the surface of the wave normals becomes a singular point resembling a navel (see Fig. 6.15). v_r or the directions of the associated Poynting vectors are thus radiating out in the form of a cone from around the "navel". Since the Poynting vectors represent the true EM energy propagation, the detector will see a cone of light. More details can be found in Born and Wolf.

6.6 Physical Discussion

It is easy to incorrectly make the following statements. Since for each direction $\hat{k}$ inside an anisotropic medium, there are in general two solutions D_1 and D_2; these two fields should each be associated with a wave vector k_1 and k_2. Thus, one has a pair of solutions (k_1, D_1) and (k_2, D_2). Because of this, one would deduce that along k_1, there can be a pair of solutions again, and so is the case for k_2. And this would propagate indefinitely so that there would be many k_n's $(n = 1, 2, 3, \ldots)$ branching out from the original k. The same question applies to the Poynting vector S (or t), which has two solutions (t_1, E_1) and (t_2, E_2). One would then say that t_1 and t_2 would branch out in the same way and so on until there are many t'_ns $(n = 1,2,3)$. We explain why such statements are incorrect.

When one says that the equation of the wave normals gives two solutions (k_1, D_1) and (k_2, D_2) for a given $\hat{k}$, it means that the anisotropic medium can support two plane waves of wave vectors k_1 and k_2 in the same direction as k. Each of the plane wavefronts contains D_1 and D_2, respectively.

In other words, when a plane EM wave spanning the entire space of the anisotropic medium interacts with the material inside, the electric field $\boldsymbol{E}(\boldsymbol{r},t)$ at every point of the wavefront will induce in the material a polarization that re-radiates according to the Maxwell equations. The solution of Maxwell's equations at each such point gives rise to a surface of wave normals. In a given direction $\hat{\boldsymbol{k}}$, the combination of all the reradiated wavefronts from all the re-radiating "point" sources on the initial wavefront gives rise to two total re-radiated wavefronts of wave vectors $\boldsymbol{k}_1$, and $\boldsymbol{k}_2$, i.e., each wavefront represents one of the following solutions:

$$\boldsymbol{D}_i = \boldsymbol{D}_{i0} \exp[i(\omega t - \boldsymbol{k}_i \cdot \boldsymbol{r})], i = 1, 2. \qquad (6.99)$$

Thus, for each direction $\hat{\boldsymbol{k}}$, the medium supports only two waves spanning the whole medium.

Similar argument can be applied to the ray surface, the ray direction $\boldsymbol{t}$, and the two wavefronts:

$$\boldsymbol{E}_i = \boldsymbol{E}_{i0} \exp[i(\omega t - \boldsymbol{k}_i \cdot \boldsymbol{r})], i = 1, 2. \qquad (6.100)$$

The above discussion applies to the hypothetical situation in which a plane wave with an arbitrary wave vector $\boldsymbol{k}$ is assumed to exist inside the anisotropic medium. In practice, we need to send such a wave from outside the medium. This coupling from outside encounters a restriction to the propagation, namely, the boundary condition at the interface between the incident and transmitted media. If the incident medium is an isotropic medium, the transmission into the anisotropic medium leads to the phenomenon of double refraction. This means that the incident plane wave excites the material across the whole interface. The radiation re-radiated by all the "point" sources at the anisotropic side of the interface satisfies both the solutions given by the surfaces of wave normals and the boundary condition (the tangential E's and H's are continuous). The result of this is that the anisotropic medium supports in general two plane waves spanning the whole medium; the wave vectors are $\boldsymbol{k}_1$ and $\boldsymbol{k}_2$. But now $\boldsymbol{k}_1$ and $\boldsymbol{k}_2$ each satisfy a different Snell's law. That means each of the two waves will propagate with a different velocity in different directions. The electric fields $\boldsymbol{E}_1$ and $\boldsymbol{E}_2$ associated with the two refracted waves are orthogonal according to Section 6.3.

Before ending this chapter, here is a question to the students: In practical applications, one often uses a uniaxial crystal. Why not a biaxial crystal?

Chapter 7

Polarization, Its Manipulation and Jones Vectors

So far, we have only talked about plane electromagnetic (EM) waves, which have linear (or plane) polarizations. That is to say, the electric vector has a fixed orientation in space as the wave propagates. In fact, ideal natural sources are assumed to randomly emit EM wave trains that are polarized linearly in all possible directions transverse to the direction of propagation.

It is only after the emission that different waves combine in different ways so that the resultant polarization becomes complicated (circular or elliptic). Such combination is based on the principle of superposition of waves, without which the nature of polarization would certainly be completely different. We shall review in this chapter all states of polarization before talking about the application of optical anisotropy to produce different states of polarization and the matrix mathematics for the propagation of polarized EM waves in anisotropic material.

7.1 Superposition of EM Waves

In what follows, all waves are assumed to travel in the z-direction. For any plane or linearly polarized monochromatic EM wave, one can choose a laboratory coordinate system to describe the electric field vector $\boldsymbol{E}(\boldsymbol{r}, t)$ associated with the wave. Thus, in this coordinate system, one can decompose $\boldsymbol{E}(\boldsymbol{r}, t)$ into $\boldsymbol{E}_x(\boldsymbol{r}, t)$ and $\boldsymbol{E}_y(\boldsymbol{r}, t)$ in the x- and y-directions of the laboratory coordinate system.

The x and y directions are arbitrary, say, horizontal and vertical. Together with the z-axis (the propagation axis), they form the normal right-handed Cartesian coordinate system. If there is another EM wave of the same frequency propagating in the same laboratory coordinate system, it too can be decomposed in the same way so that the x- and y-components can be superimposed, giving a resultant component in the x-direction and a resultant component in the y-direction. In general, there can be many monochromatic plane waves of the same frequency, each of which is decomposed into an x- and y-component. One can then superimpose all x- and y-components separately, resulting in two resultant components of traveling waves:

$$\boldsymbol{E}_x = \hat{\boldsymbol{i}} E_{ox} \cos(\omega t - kz + \epsilon_x), \tag{7.1}$$

$$\boldsymbol{E}_y = \hat{\boldsymbol{j}} E_{oy} \cos(\omega t - kz + \epsilon_y). \tag{7.2}$$

ϵ_x, ϵ_y are the phases. We rewrite these two expressions as:

$$\boldsymbol{E}_x = \hat{\boldsymbol{i}} E_{ox} \cos(\omega t - kz) \tag{7.3}$$

$$\boldsymbol{E}_y = \hat{\boldsymbol{j}} E_{oy} \cos(\omega t - kz + \epsilon), \tag{7.4}$$

where

$$\epsilon = \epsilon_x + \epsilon_y. \tag{7.5}$$

This is valid because we will be dealing with the phase difference ϵ only.

Definition:

$$E_y \text{ leads } E_x \text{ by } \epsilon \text{ if } \epsilon > 0$$

and

$$E_y \text{ lags behind } E_x \text{ by } \epsilon \text{ if } \epsilon < 0.$$

The above definition can be seen by inspecting Eqs. (7.3) and (7.4). If $\epsilon > 0$ and at a fixed position in space, E_y will become zero (for example) at $\frac{\epsilon}{\omega}$ seconds earlier than E_x (Fig. 7.1(a)). A similar explanation applies to phase lag if $\epsilon < 0$. Such phase lag or lead is easier to see by considering the complex representation (Fig. 7.1(b) and (c)). Equations (7.3) and (7.4) can now be vectorially added so as to obtain a resultant electric field. The state (of polarization) of the resultant electric field depends on ϵ. We discuss this in the following sections.

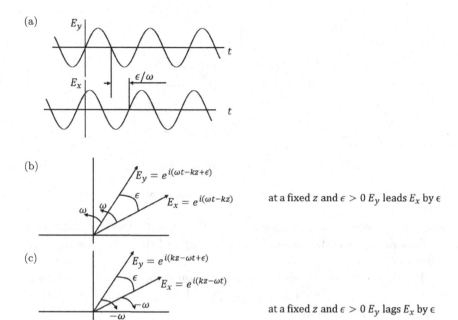

Fig. 7.1. Two sinusoidal oscillations with a phase difference: definition of phase lead and phase lag.

7.2 Linear Polarization

$$\text{Condition} \quad \epsilon = \begin{cases} 2n\pi \\ (2n+1)\pi \end{cases} (n = 0, 1, 2, 3 \ldots \ldots) \qquad (7.6)$$

Thus

$$E_{ox} \neq E_{oy},$$

$$E_y = E_{oy} \begin{cases} \cos(\omega t - kz + 2n\pi), \\ \cos(\omega t - kz + (2n+1)\pi), \end{cases} \qquad (7.7)$$

$$= \pm E_{oy} \cos(\omega t - kz),$$

$$\therefore \boldsymbol{E} = \boldsymbol{E}_x + \boldsymbol{E}_y,$$

$$= \boldsymbol{E}_{ox} \cos(\omega t - kz) \pm \boldsymbol{E}_{oy} \cos(\omega t - kz), \qquad (7.8)$$

$$= (\boldsymbol{E}_{ox} \pm \boldsymbol{E}_{oy}) \cos(\omega t - kz).$$

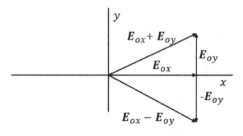

$$\epsilon = \begin{cases} 2n\pi \\ (2n+1)\pi \end{cases} \rightarrow \text{linear polarization}$$

Fig. 7.2. Definition of linear polarization.

Each of the vector amplitudes $(\boldsymbol{E}_{ox} \pm \boldsymbol{E}_{oy})$ has a fixed orientation in the laboratory coordinate system (Fig. 7.2) and they are thus linearly polarized.

Traditionally, one defines linear polarization as $\epsilon = \pi$, i.e., setting $n = 0$ in Eq. (7.6). This is because $2n\pi$ does not change the cosine or exponential function.

7.3 Circular Polarization

Conditions:

$$\epsilon = 2n\pi \pm \frac{\pi}{2}(n = 0, 1, 2, \ldots) \tag{7.9}$$

and

$$E_{ox} = E_{oy} = E_o.$$

Thus,

$$E_y = E_o \cos(\omega t - kz + 2m \pm \frac{\pi}{2}),$$
$$= \mp E_o \sin(\omega t - kz). \tag{7.10}$$

Also,

$$E_x = E_o \cos(\omega t - kz). \tag{7.11}$$

Now, the resultant vector traces out a circle in the $x - y$ plane. To see this, we add:

$$E_x^2 + E_y^2 = E_o^2[\sin^2(\omega t - kz) + \cos^2(\omega t - kz)] = E_o^2, \tag{7.12'}$$

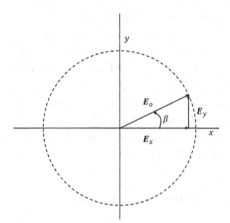

Fig. 7.3.   Definition of circular polarization if one looks into the direction of propagation at a fixed position z.

i.e., the x and y components together form a circle of radius E_o, or the tip of E_o traces out a circle in the $x - y$ plane. Now at a certain time t, the x and y components are given by Eqs. (7.10) and (7.11) (see Fig. 7.3).

    Thus,

$$\tan \beta \equiv \frac{E}{E_x} = \mp \tan(\omega t - kz), \qquad (7.12)$$

or

$$\beta = \mp(\omega t - kz), \qquad (7.13)$$

$$\therefore \frac{d\beta}{dt} = \mp \omega \text{ for } \epsilon = 2n\pi \pm \pi/2 \text{ and } z = \text{constant.} \qquad (7.14)$$

This means that β changes in time at a fixed position z (Fig. 7.3), i.e., the resultant electric field E_o rotates at the rate of $\pm \omega$ in the counterclockwise or clockwise direction for $(\epsilon = 2n\pi - \pi/2)$ and $(\epsilon = 2n\pi + \pi/2)$, respectively, if one looks into the direction of the light beam at a fixed position z. The above senses of rotation can also be called left-handed (counter-clockwise) and right-handed (clockwise) because they follow the left and right-hand "rules" of rotation, respectively. All these are confusing. Worse, the above definitions of the senses of rotation will be reversed if one changes the phase of the wave from $(\omega t - kz)$ to $(kz - \omega t)$. (A more detailed comment follows the section on elliptic polarization.) Also, as words like "clockwise" and "counter-clockwise" will sooner or later go out of fashion since most mechanical clocks and watches will probably be replaced by digital clocks or watches, younger generations will find it difficult to use such words. As such, a symbol will be used to represent the sense of rotation and state of polarization. They are defined as follows (assuming we always look into the light beam):

Circular polarizations are simply represented by two circles with opposite senses of rotation: ↺ and ↻

Again, traditionally, as in the case of linear polarization, one defines circular polarization as when $\epsilon = \pm \pi/2$, i.e., $n = 0$.

7.4   General or Elliptic Polarization

For all other values of ϵ (other than $2n\pi$, $(2n+1)\pi$ and $2n\pi \pm \pi/2$) and arbitrary values of E_{ox} and E_{oy}, the state of polarization of the resultant electric field is elliptical. This can be seen by calculating the following:

$$\left(\frac{E_x}{E_{ox}}\right)^2 + \left(\frac{E_y}{E_{oy}}\right)^2 - 2\left(\frac{E_x}{E_{ox}}\right)\left(\frac{E_y}{E_{oy}}\right)\cos\epsilon \qquad (7.15)$$

$$= \cdots (\underline{\text{exercise}} \text{ for the reader using Eqs. (7.3 and 7.4)})$$

$$= \sin^2\epsilon$$

Equation (7.15) is an ellipse. One can make a proper rotation of the $x - y$ axes such that Equation (7.15) becomes a familiar equation of an ellipse.

$$\left(\frac{E'_x}{E'_{ox}}\right)^2 + \left(\frac{E'_y}{E'_{oy}}\right)^2 = 1. \qquad (7.16)$$

The "′" (prime) means the fields are now components in the rotated x' y' coordinate system. The details of the rotation of coordinates are given below:

$$E_{ox} = x_o \, E'_{ox} = x'_o,$$
$$E_{oy} = y_o \, E'_{oy} = y'_o, \qquad (7.17)$$
$$E_x = x \, E'_x = x',$$
$$E_y = y \, E'_y = y'.$$

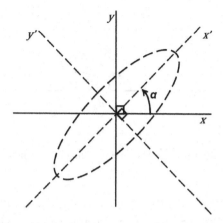

Fig. 7.4.   Definition of elliptic polarization.

We do the coordinate rotation:

$$x = x'\cos\alpha - y'\sin\alpha, \tag{7.18}$$

$$y = x'\sin\alpha + y'\cos\alpha. \tag{7.19}$$

Substitute Eqs. (7.18) and (7.19) into (7.15) and requiring that Eq. (7.16) be valid (see Fig. 7.4), one obtains (<u>exercise</u> for the reader):

$$\tan 2\alpha = \frac{2x_o y_o}{x_o^2 - y_o^2}\cos\epsilon \tag{7.20}$$

and

$$x_o'^2 = \frac{x_o^2 y_o^2 \sin^2\epsilon}{y_o^2\cos^2\alpha + x_o^2\sin^2\alpha - 2x_o y_o \sin\alpha\cos\alpha\cos\epsilon}, \tag{7.21}$$

$$y_o'^2 = \frac{x_o^2 y_o^2 \sin^2\epsilon}{y_o^2\sin^2\alpha + x_o^2\cos^2\alpha + 2x_o y_o \sin\alpha\cos\alpha\cos\epsilon}. \tag{7.22}$$

After this rotation, the "normal" look of the ellipse in the $x'y'$ coordinates is shown in Fig. 7.4. The sense of rotation of the tip of the resultant electric field around the ellipse can be determined as follows. In the $x'y'$ frame, we define β as (see Fig. 7.5):

$$\tan\beta = \frac{y'}{x'} \equiv \frac{E_y'}{E_x'} \quad \text{(by Eq. (7.17))},$$

$$= \frac{E_{oy}'\cos(\omega t - kz + \epsilon)}{E_{ox}'\cos(\omega t - kz)}. \tag{7.23}$$

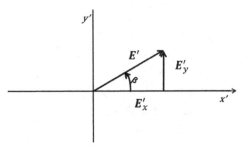

Fig. 7.5. Relationship between the orthogonal electric fields of an elliptically polarized wave.

Note: Since from Eq. (7.20), $\alpha = \alpha(x_o, y_o, \epsilon)$ only, the rotation of the $x - y$ coordinates by α does not affect $(\omega t - kz)$. Hence, we use the initial cosine terms in Eq. (7.23). After some simplification:

$$\tan\beta = \frac{E'_{oy}}{E'_{ox}}\{\cos\epsilon - \tan(\omega t - kz)\sin\epsilon\}. \tag{7.24}$$

Differentiate with respect to time t at a fixed position z:

$$\sec^2\beta\frac{\mathrm{d}\beta}{\mathrm{d}t} = \frac{E'_{oy}}{E'_{ox}}\{-[\sin\epsilon\sec^2(\omega t - kz)]\omega\}.$$

Hence:

$$\frac{\mathrm{d}\beta}{\mathrm{d}t} = -\frac{E'_{oy}}{E'_{ox}}\omega\frac{\sec^2(\omega t - kz)}{\sec^2\beta}\sin\epsilon. \tag{7.25}$$

$\therefore$ The first three parts on the right hand side are positive, we have:

$$\frac{\mathrm{d}\beta}{\mathrm{d}t}\alpha - \sin\epsilon. \tag{7.26}$$

We have the following results:

For an observer at a fixed position z looking into the light beam:

$$(-\sin\epsilon) > 0 \text{ means } \frac{\mathrm{d}\beta}{\mathrm{d}t} > 0 \text{ leading to } \circlearrowleft$$

$$(-\sin\epsilon) < 0 \text{ means } \frac{\mathrm{d}\beta}{\mathrm{d}t} < 0 \text{ leading to } \circlearrowright \tag{7.27a}$$

$$(-\sin\epsilon) = 0 \text{ means } \frac{\mathrm{d}\beta}{\mathrm{d}t} = 0 \text{ leading to } \nearrow$$

The symbol of an ellipse represents the state of elliptic polarization whose sense of rotation is indicated. The last line for linear

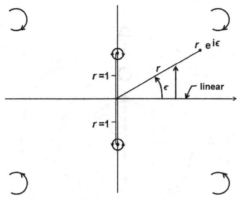

Fig. 7.6. Complex plane defining different polarization states of an EM wave where $r \equiv y_0/x_0 \equiv E_{0y}/E_{0x}$.

polarization (double-ended arrow inclined at an arbitrary angle) is true because $\left(\frac{d\beta}{dt} = 0\right)$ leads to β =constant, i.e., the resultant electric vector has a fixed orientation in space as the wave propagates. Furthermore, we add, for completeness, that for $E_{ox} = E_{oy}$ and $-\sin\epsilon = \pm 1$, one obtains circular polarizations (Exercise for the reader who will prove that $(E_{oy}/E_{ox} = E'_{oy}/E'_{ox})$).

To summarize, one can determine the orientation of the ellipse using Eq. (7.20), the major and minor axes of the ellipse by Eqs. 7.21) and (7.22), and the sense of rotation by Eq. (7.26). All these four equations depend only on ϵ and $r \equiv y_0/x_0$. Thus, the state of polarization and its size, orientation and sense of rotation are all completely determined if one knows r and ϵ. Because of this, one can represent all the states and their sizes and senses of rotation in a complex plane using the vector (Fig. 7.6)

$$re^{i\epsilon}. \tag{7.27b}$$

Each point in this complex plane corresponds to one polarization. We shall analyze this.

From Eqs. (7.27a) and Fig. 7.6, we see that the upper half of the complex plane contains all polarization states whose sense of rotation is $\curvearrowright$ ($\sin\epsilon > 0$) while the lower half is $\curvearrowright$ ($\sin\epsilon < 0$). The horizontal axis ($\sin\epsilon = 0$) contains all the states of linear polarization with different orientations. There are only two points ($r = 1$) on the vertical axis that give the two states of circular polarization (along

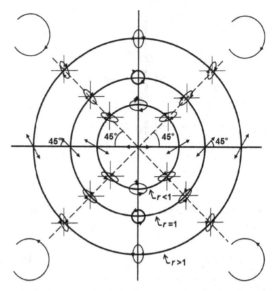

Fig. 7.7. Polarization chart in the complex plane of polarization.

the vertical axis, $\sin \epsilon = \pm 1$, plus the condition $r = 1$, one obtains circular polarization). Besides the above special cases, all the other points in the complex plane give elliptic polarizations.

Figure 7.7 shows polarization states in the complex plane around three circles, $r < 1$, $r = 1$ and $r > 1$. The reader should try (exercise) to prove that the orientations of the elliptic polarizations and linear polarizations are indeed of what is shown in Fig. 7.7.

How to use the polarization chart of Fig. 7.7.

To use the chart, one needs to know r and ϵ. We can then locate the point $re^{i\epsilon}$ in the complex plane and determine quickly and qualitatively what the orientation and sense of rotation of the polarization state are. Only when the points lie along the horizontal axis will the polarization be linear, and circular polarizations occur only at two points on the vertical axis ($\epsilon = \pm \pi/2$), namely, $r = \pm 1$ (i.e., $E_{ox} = E_{oy}$).

7.5   Some Comments on the Sense of Rotation of Circular and Elliptical Polarization

The sense of rotation (left-handed or right-handed, clockwise or counterclockwise) of circular and elliptical polarizations depends on two definitions.

(a) The position of the observer: Whether he looks into the light beam or towards the beam's direction of propagation. (Most people, including us, choose the former. If one chooses the latter definition, all the senses of rotation will be reversed.)

(b) The mathematical description of the wave propagation: Whether the propagating wave function is described by $f(\omega t - kz)$ or $f(kz - \omega t)$. (For example, $\cos(\omega t - kz)$ or $\cos(kz - \omega t)$.) (Note: kz can be replaced by $\mathbf{k} \cdot \mathbf{r}$ Both usages are equally popular. The sense of rotation of the polarization resulting from one definition of the wave function is opposite to that resulting from the other. This can easily be seen in Eq. (7.13) by changing $(\omega t - kz)$ to $(kz - \omega t)$ and the sign of $d\beta/dt$ in Eq. (7.14) is reversed.

Such a relative sense of rotation of the polarization (circular or elliptical) might lead to confusion. Thus, when talking about the sense of rotation, it is wise to first define the two "frames of reference" mentioned in (a) and (b). Comparison between results will then be consistent.

One might thus ask the following question. If the sense of rotation is so relative, are there really circular and elliptical polarizations that rotate in definite senses? The answer is "yes". Circular and elliptical polarizations do have definite senses of rotation. Whether we call one right-handed or left-handed (counter-clockwise or clockwise) depends strictly on our choice of definition. It is similar to watching a spinning top. The sense of rotation is definitely there but whether it is described as right or left, clockwise or anticlockwise, etc., is a matter of choice.

7.6 Anisotropic Material as Polarizer

Using the phenomenon of double refraction in an anisotropic material described in Chapter 6, one can devise ways to isolate linearly polarized light from an EM radiation with random polarizations. One way is to use a *Nicol prism* (Fig. 7.8(a)). The Nicol prism is essentially an assembly of two anisotropic prisms stuck together by transparent cement. Usually, the prisms are calcite, while the cement is Canada balsam. Calcite is a uniaxial crystal. The optic axes of the two prisms are parallel to each other, as shown in Fig. 7.8(a).

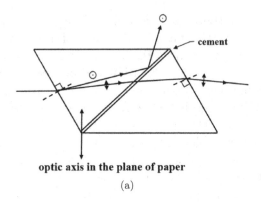

optic axis in the plane of paper

(a)

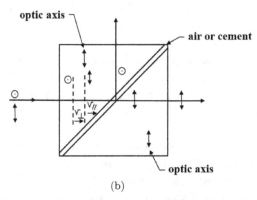

(b)

Fig. 7.8.   (a) A Nicol prism and (b) a Glan prism.

A randomly polarized EM wave enters the prism at a non-zero angle of incidence. Double refraction produces two beams of orthogonal linear polarizations — one (ordinary (o) wave) perpendicular to the principal plane containing the optic axis, the other (extraordinary (e) wave) in the principal plane. (Thus, the polarization of the o-wave is always perpendicular to the optic axes.) When these two beams (waves) of different velocities (hence different indices of refraction n_o, n_e) arrive at the interface of the two prisms, the o-wave is totally internally reflected, and the extraordinary wave passes through into the second prism. This requires that the cement's index of refraction n_c be between n_o and n_e, i.e., $n_e < n_c < n_o$. No further double refraction or double wavefronts are generated in the second medium because the polarization of the extraordinary wave is now in the principal plane containing the optic axis (see Section 6.3).

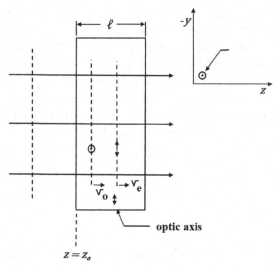

Fig. 7.9. A general wave plate.

Different variations of such types of prism exist. They are used as beam splitters of laser beams inside and outside laser cavities (*Glan air* or *Glan Foucault prisms, Glan-Thomson prisms*, again made of calcite, see Fig. 7.8(b)). A laser beam of arbitrary polarization enters the surface normally. The two waves (*o* and *e*) thus created inside the prism propagate in the same direction but with different velocities and orthogonal polarizations. As in the case of the Nicol prism, one (*o*-wave) is rejected, and one (*e*-wave) is transmitted. If the laser beam's polarization is parallel to that of the *o*-wave, it will be rejected; if it is parallel to that of the *e*-wave, it will be transmitted.

7.7 Wave Plates and Wave Retarders

Using an anisotropic material, one can transform one state of polarization of an EM wave into another. This is usually accomplished by a wave plate with parallel incident and exit surfaces. As shown in Fig. 7.9, a plane monochromatic EM wave of any polarization enters a plate of anisotropic material (usually uniaxial crystals) at normal incidence. The optic axis is perpendicular to z as shown, assuming the material to be a uniaxial crystal. Thus, the two waves (ordinary and extraordinary) generated in the medium will both travel in the same direction but at different velocities v_o and v_e corresponding to two indices n_o and n_e. These two waves are initially in phase at the incident interface. After propagating through the plate of thickness ℓ, a phase difference ϵ is developed between the two waves. Let x and y represent the polarization direction of the ordinary wave and extraordinary wave, respectively. Then the incident (plane) wave can be decomposed into components along the x and y-axes.

$$E_x = E_{ox} \cos(\omega t - kz),$$

$$E_y = E_{oy} \cos(\omega t - kz),$$

where

$$k = \frac{\omega}{c},$$

assuming that the incident medium is vacuum. It could also be an isotropic medium. In this case, c becomes v, the phase velocity. At the exit of the wave plate,

$$E_x = E_{ox} \cos(\omega t - kz - k_o\ell), \qquad (7.28)$$

$$E_y = E_{oy} \cos(\omega t - kz - k_e\ell), \qquad (7.29)$$

where

$$k_o = \frac{\omega}{c} n_o, \qquad (7.30)$$

$$k_e = \frac{\omega}{c} n_e. \qquad (7.31)$$

o: ordinary e: extraordinary

Thus, the phase difference between E_x and E_y is, by Eq. (7.5):

$$\epsilon = -(k_e - k_o)\ell$$

or

$$\epsilon = \frac{\omega}{c} \ell (n_o - n_e). \qquad (7.32)$$

Rewriting Eqs. (7.28) and (7.29) in the form of Eqs. (7.3) and (7.4), we have:

$$E_x = E_{ox} \cos(\omega t - kz), \qquad (7.33)$$

$$E_y = E_{oy} \cos(\omega t - kz + \epsilon), \qquad (7.34)$$

which are identical to Eqs. (7.3) and (7.4). At the exit of the plate, the two waves will exit normally to the surface because the entrance and exit surfaces are parallel. They will recombine vectorially, producing different states of polarization depending on the value of ϵ given by Eq. (7.32). We look at some common wave plates:

(a) *Quarter wave plate* ($\lambda/4$ plate)

$$\text{Conditions}: \epsilon = 2n\pi \pm \pi/2 \tag{7.35}$$

These are the conditions for the generation of circular polarizations (see Section 7.3), if, in addition,

$$E_{ox} = E_{oy}. \tag{7.36}$$

Equation (7.35) means that (using Eq. (7.32)):

$$\frac{\omega}{c}\ell(n_o - n_e) = 2n\pi \pm \pi/2,$$

$$\ell(n_o - n_e) = \frac{c}{\omega}(2n\pi \pm \pi/2)$$

or

$$\ell(n_o - n_e) = \frac{\lambda}{2\pi}(2n\pi \pm \pi/2)$$

$$= \left(n \pm \frac{1}{4}\right)\lambda, \quad n = 0, 1, 2, \ldots \tag{7.37}$$

i.e., the optical path difference between the o and e waves in traversing the quarter wave plate is $n\lambda$ plus or minus a quarter of a wavelength (λ) — "minus" sign for ⟳, "plus" for ⟲. Equation (7.36) means that the incident polarization should be linear and make an angle of 45° with the x - and y-axes so that $E_{ox} = E_{oy}$.

Because the wave plate is passive, and the o and e waves propagate in the same direction for the incident wave normal to the surface, the reverse is also true, i.e., if the incident wave is circularly polarized, the transmitted wave will be linearly polarized. To see this, we decompose the incident wave into:

$$E_x = E_o \cos(\omega t - kz), \tag{7.38}$$

$$E_y = E_o \cos\left(\omega t - kz + 2n\pi \pm \frac{\pi}{2}\right), \tag{7.39}$$

which is the definition of circular polarization. After traversing the plate, an additional phase difference of $(2n\pi \pm \pi/2)$ is added to the phase of E_y.

$$E_y = E_o \cos(\omega t - kz + 4n\pi \pm \pi),$$
$$= -E_o \cos(\omega t - kz). \tag{7.40}$$

Combining Eqs. (7.38) and (7.40):

$$\boldsymbol{E} = \boldsymbol{E}_x + \boldsymbol{E}_y,$$
$$= \boldsymbol{E}_o(\hat{i} - \hat{j}) \cos(\omega t - kz), \tag{7.41}$$

which is a linear polarization making an angle of 45° with the x and y axes.

Summary: For a quarter-wave plate: linear polarization in, circular out and vice versa.

(b) *Half-wave plate* ($\lambda/2$ plate)

Condition:

$$\epsilon = (2n+1)\pi \tag{7.42}$$

This is the condition of linear polarization (Section 7.2), i.e., the transmitted wave is linearly polarized. This is not a sufficient statement. Let us look more closely. Equation (7.42) into (7.32) gives:

$$(n_o - n_e)\ell = \frac{\lambda}{2\pi} \cdot (2n+1)\pi = n\lambda + \frac{\lambda}{2}, \tag{7.43}$$

i.e., the optical path difference between the o and e waves in traversing the half-wave plate is $n\lambda$ plus half of a wavelength. If the incident wave is a linearly polarized plane wave with components:

$$E_x = E_{ox} \cos(\omega t - kz), \tag{7.44}$$
$$E_y = E_{oy} \cos(\omega t - kz). \tag{7.45}$$

Then the transmitted wave components (relative to each other) become:

$$E_x = E_{ox} \cos(\omega t - kz), \tag{7.46}$$
$$E_y = E_{oy} \cos(\omega t - kz + (2n+1)\pi), \tag{7.47}$$
$$\text{i.e.,} \quad E_y = -E_{oy} \cos(\omega t - kz). \tag{7.48}$$

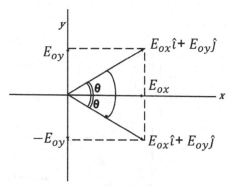

Fig. 7.10.   Effect of a half-wave plate: polarization rotation.

Combining Eqs. (7.46) and (7.48), we see that the transmitted electric vector becomes:

$$\boldsymbol{E}_t = (E_{ox}\hat{i} - E_{oy}\hat{j})\cos(\omega t - kz). \tag{7.49}$$

Compare with the incident vector (Eqs. (7.44) and (7.45):

$$\boldsymbol{E}_i = (E_{ox}\hat{i} + E_{oy}\hat{j})\cos(\omega t - kz). \tag{7.50}$$

We see that $\boldsymbol{E}_i$ has been rotated by an angle 2θ in traversing the half-wave plate (Fig. 7.10) where θ is the angle between $\boldsymbol{E}_i$ and $\hat{I}$. In practice, θ is often set at 45°. Then the effect of the half-wave plate is to rotate the incident linear polarization by 90°. However, θ can be any value so that $0 \le 2\theta \le \pi$. This means that a $\lambda/2$ plate acts as a polarization rotator.

(c) *A general wave plate*

For a general value of ϵ, the resultant transmitted wave will be elliptically polarized (Section 7.4), if the incident wave is linearly polarized.

(d) *A wave retarder* (Babinet compensator)

In contrast to a wave plate that introduces a fixed phase shift, a wave retarder is an anisotropic device whose effect is to introduce a continuously variable relative phase shift between the x and y components of an incident wave. The central design of a wave retarder is to stack two anisotropic wedges on top of each other (Fig. 7.11) with their optic axes perpendicular to each other assuming uniaxial crystals. This is called a *Babinet compensator*. A linearly polarized

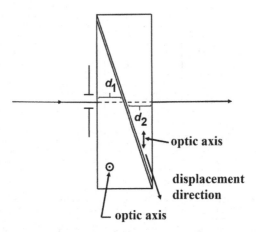

Fig. 7.11. A wave retarder: Babinet compensator.

laser beam whose electric field is $\boldsymbol{E}_i$ enters the surface at normal incidence into the retarder.

Inside the left wedge, there are two waves (o and e) propagating in the same direction. The polarization of the o-wave is perpendicular to the optic axis while the e-wave is parallel. The relative phase shift between the o and e waves after traversing the left wedge through the thickness d_1 is (by Eq. (7.32)):

$$\epsilon_1 = \frac{\omega}{c} d_1 (n_o - n_e). \tag{7.51}$$

After entering the right wedge, the o-wave becomes an e-wave while the e-wave becomes an o-wave because the optic axis is now parallel to the polarization of the o-wave inside the left wedge. The relative phase difference in traversing the thickness d_2 of the right wedge is (by Eq. (7.32)):

$$\epsilon_2 = \frac{\omega}{c} d_2 (n_e - n_o). \tag{7.52}$$

After traversing the whole retarder, the total relative phase shift is:

$$\epsilon = \frac{\omega}{c} d_1 (n_o - n_e) + \frac{\omega}{c} d_2 (n_e - n_o),$$

$$\epsilon = \frac{\omega}{c} (n_o - n_e)(d_1 - d_2). \tag{7.53}$$

By sliding one wedge with respect to the other along the contact surface, $(d_1 - d_2)$ can be varied continuously, and by Eq. (7.53), the total relative phase shift ϵ is also varied continuously.

Different variations of wave retarders have been designed, and the reader is referred to an optics book for more detail (see, for example, (Hecht, 1987)).

7.8 Jones Vectors

The previous section analyzes from the physical point of view the change in the polarization state of an EM wave that propagates through wave plates and retarders. The procedure is rather time-consuming, especially if one needs to analyze the propagation through a series of anisotropic media. Using the Jones vector, a straightforward matrix calculation can be applied to solve the above problem "quickly".

Definition: *Jones vector*

We first decompose the electric field vector of an EM wave into the x and y components in a laboratory reference frame. For convenience, we call x "horizontal" and y "vertical". At a fixed position z along the propagation direction:

$$\boldsymbol{E}(\boldsymbol{r}, t) = \boldsymbol{E}_x(\boldsymbol{r}, t)\hat{i} + E_y(\boldsymbol{r}, t)\hat{j}, \qquad (7.54)$$

$$\text{Jones vector } \boldsymbol{J} \equiv \begin{bmatrix} \boldsymbol{E}_x(\boldsymbol{r}, t) \\ \boldsymbol{E}_y(\boldsymbol{r}, t) \end{bmatrix}. \qquad (7.55)$$

As we have seen in the previous section, when the EM wave propagates through wave plates, the original phase of $\boldsymbol{E}$, i.e., $(\omega t - kz)$, was not affected. Only phase shifts were added to the original phase. Hence, we can neglect the phase $(\omega t - kz)$ and write Eq. (7.55) as:

$$\boldsymbol{J} = \begin{bmatrix} E_{ox}e^{i\varphi_x} \\ E_{oy}e^{i\varphi_y} \end{bmatrix}, \qquad (7.56)$$

where exponential notation, rather than cosine function, is required. Therefore, the Jones vector is a complex vector in an abstract space.

To obtain the real field, one has to calculate, for example:

$$E_x = Re\{J_x e^{i(\omega t - kz)}\} = Re\{E_{ox} e^{i(\omega t - kz + \varphi_X)}\},$$

$$= E_{ox} \cos(\omega t - kz + \varphi_X). \tag{7.57}$$

Often, one is only interested in the polarization state of the wave after propagation through anisotropic media. Only a relative phase difference is needed (see the previous section). Equation (7.56) thus becomes:

$$J = e^{i\varphi_X} \begin{bmatrix} E_{ox} \\ E_{oy} e^{i(\varphi_y - \varphi_x)} \end{bmatrix}.$$

Omitting $e^{i\varphi_X}$, because it is a common phase factor, we have:

$$J = \begin{bmatrix} E_{ox} \\ E_{oy} e^{i\epsilon} \end{bmatrix}, \tag{7.58}$$

where

$$\epsilon \equiv \varphi_y - \varphi_x. \tag{7.58'}$$

(Here, as in Sections 7.2 to 7.4, we use the x-component as the reference and consider the phase difference of y with respect to x-components.) We can normalize the amplitude of the Jones vector and retain the *relative* phase difference. To normalize Eq. (7.58), let:

$$J_N = \begin{bmatrix} E_{Nx} \\ E_{Ny} \end{bmatrix}, \tag{7.59}$$

and require

$$J_N \cdot J_N^* = 1, \tag{7.60}$$

where the asterisk "*" denotes the complex conjugate. This means:

$$E_{Nx} \cdot E_{Nx}^* + E_{Ny} \cdot E_{Ny}^* = 1 \tag{7.61}$$

and

$$E_{Nx} = \frac{E_{ox}}{(J \cdot J^*)^{1/2}}, \tag{7.62}$$

$$E_{Ny} = \frac{E_{oy} e^{i\epsilon}}{(J \cdot J^*)^{1/2}}. \tag{7.63}$$

Note (1):

$$(J \cdot J^*) = E_{ox} \cdot E_{ox}^* + E_{oy} \cdot E_{oy}^*,$$

$$= E_{ox}^2 + E_{oy}^2, \quad \text{if } E_{ox}, E_{oy} \text{ are real} \tag{7.64}$$

Note (2): The definition of the Jones vector (Eq. (7.57)) is *relative*. In Eq. (7.57), we used the x-component as our reference, similar to the definition of polarization, Eqs. (7.3) and (7.4). One can also define the Jones vector using the y-component as the reference. In such a case:

$$J = \begin{bmatrix} E_{ox} e^i \epsilon' \\ E_{oy} \end{bmatrix}, \quad \epsilon' \equiv \varphi_x - \varphi_y.$$

We shall stick to the first definition (7.57).

Example 1: A general *linear polarization* making an angle φ with the x-axis. We first determine the general Jones vector (Eq. (7.56)). Since the x and y components of such a linear polarization are in phase,

i.e., $\quad \boldsymbol{E} = \boldsymbol{E}_o \cos(\omega t - kz),$

$$= [(E_o \cos \psi)\hat{i} + (E_o \sin \psi)\hat{j}] \cos(\omega t - kz).$$

The Jones vector is thus:

$$J = \begin{bmatrix} E_o \cos\psi \\ E_o \sin\psi \end{bmatrix}. \tag{7.65}$$

To normalize, we calculate:

$$\boldsymbol{J} \cdot \boldsymbol{J}^\star = E_o^2(\cos^2\psi + \sin^2\psi) = E_o^2,$$

$$E_{Nx} = \frac{E_o \cos \psi}{E_o} = \cos \psi,$$

$$E_{Ny} = \frac{E_o \cos \psi}{E_o} = \sin \psi,$$

$$\therefore \quad \boldsymbol{J}_N = \begin{bmatrix} \cos\psi \\ \sin\psi \end{bmatrix}. \tag{7.66}$$

If $\psi = 0$, we have horizontal polarization, and

$$\boldsymbol{J}_{Nh} = \begin{bmatrix} 1 \\ 0 \end{bmatrix}. \tag{7.67}$$

If $\psi = \pi/2$, we have vertical polarization and

$$\boldsymbol{J}_{NV} = \begin{bmatrix} 0 \\ 1 \end{bmatrix}. \tag{7.68}$$

Example 2: Circular polarization. From **Section 7.3**, the field for a circular polarization is:

$$\boldsymbol{E} = E_o \cos(\omega t - kz)\hat{i} + E_o \cos\left(\omega t - kz + 2n\pi + \frac{\pi}{2}\right)\hat{j}. \quad (7.69)$$

Compare with Eq. (7.57), we have:

$$\epsilon = 2n\pi \pm \pi/2, \varphi_x = 0, E_{ox} = E_{oy} \equiv E_o.$$

Thus, substituting into Eq. (7.57):

$$\boldsymbol{J} = \begin{bmatrix} E_o \\ E_o^e \pm i\pi/2 \end{bmatrix}. \quad (7.70)$$

To normalize, we calculate:

$$\boldsymbol{J}\cdot\boldsymbol{J}^\star = E_o^2 + E_o^2 = 2E_o^2, \quad (7.71)$$

$$E_{Nx} = \frac{E_o}{\sqrt{2}E_o} = \frac{1}{\sqrt{2}}, \quad (7.72)$$

$$E_{Ny} = \frac{E_o e^{\pm i\pi/2}}{\sqrt{2}E_o} = \frac{1}{\sqrt{2}}(\pm i), \quad (7.73)$$

$$\therefore \quad \boldsymbol{J}_N = \frac{1}{\sqrt{2}}\begin{bmatrix} 1 \\ \pm i \end{bmatrix} \quad (7.74)$$

$$\text{or} \quad \boldsymbol{J}_N\left(\circlearrowright\right) = \frac{1}{\sqrt{2}}\begin{bmatrix} 1 \\ i \end{bmatrix}, \quad (7.75)$$

$$\boldsymbol{J}_N\left(\circlearrowleft\right) = \frac{1}{\sqrt{2}}\begin{bmatrix} 1 \\ -i \end{bmatrix}. \quad (7.76)$$

The senses of rotation are determined by comparing with the definitions in Section 7.3,

$$\text{i.e.,} \quad \epsilon = 2n\pi \pm \pi/2 \text{ for } \circlearrowright$$

and

$$\epsilon = 2n\pi - \pi/2 \text{ for } \circlearrowleft.$$

Example 3: *Elliptic polarization*

Equation (7.57) is the general Jones vector, i.e., it represents the most general state of polarization, which is elliptical. Normalization of this general vector:

$$J \equiv \begin{bmatrix} E_{ox} \\ E_{oy}e^{i\epsilon} \end{bmatrix} \tag{7.77}$$

gives:

$$J_N = \frac{1}{\sqrt{1+\tan^2\theta}} \begin{bmatrix} 1 \\ e^{i\epsilon}\tan\theta \end{bmatrix}, \tag{7.78}$$

where

$$\tan\theta = \frac{E_{oy}}{E_{ox}}. \tag{7.79}$$

The derivation of (7.78) is left as an <u>exercise</u> for the reader.

7.9 Propagation Through Wave Plates Using Jones Matrix Formalism

We now consider the propagation through a wave plate. The incident wave can be decomposed into E_x and E_y in the laboratory frame (say, horizontal and vertical, for convenience). The passage of the wave into the wave plate immediately induces the two waves o and e, whose polarizations are $\boldsymbol{E}_o$ and $\boldsymbol{E}_e$, respectively. This change of polarization from $(\boldsymbol{E}_x, \boldsymbol{E}_y)$ to $(\boldsymbol{E}_o, \boldsymbol{E}_e)$ is equivalent to a rotation of the coordinates (x, y) into (o, e) through an angle ψ. (Note: $\boldsymbol{E}_o \perp \boldsymbol{E}_e$, hence (o, e) forms a Cartesian coordinate) (see Fig. 7.12). The o and e polarization directions are often called *slow* and *fast* axes in the literature because of negative uniaxial materials, $v_e > v_o$. The o and e waves propagate through the wave plate of thickness l. They suffer phase shifts of $k_o\ell$ and $k_e\ell$, respectively. The relative phase shift or phase difference between the o and e waves is thus:

$$\epsilon = (k_o - k_e)\ell. \tag{7.80}$$

When the waves exit the plate, we measure them in the laboratory (x, y) frame again. This is equivalent to rotating the (o, e) frame back into the (x, y) frame (by an angle $-\psi$).

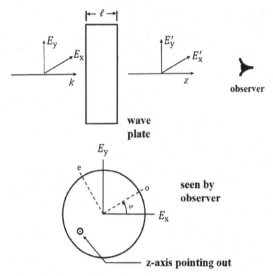

Fig. 7.12. Detailed analysis of the function of a wave plate made of uniaxial material using the Jones vector formalism.

Thus, the propagation of an EM wave through a wave plate is equivalent to three steps of operation: rotation of the (x, y) frame by an angle ψ (into the (o, e) frame), shifting of phases of the o and e waves, and rotation of the (o, e) frame by an angle $-\psi$ (back into the (x, y) frame). Rotation of coordinates can be represented by matrices. Thus, the rotation by an angle ψ is given by:

$$R(\psi) \equiv \begin{bmatrix} \cos\psi & \sin\psi \\ -\sin\psi & \cos\psi \end{bmatrix}, \tag{7.81}$$

while rotation by an angle $(-\psi)$ is:

$$R(-\psi) \equiv \begin{bmatrix} \cos\psi & -\sin\psi \\ \sin\psi & \cos\psi \end{bmatrix}. \tag{7.82}$$

Let $\begin{bmatrix} E_o \\ E_e \end{bmatrix}$ and $\begin{bmatrix} E'_o \\ E'_e \end{bmatrix}$ be the Jones vectors representing the o and e waves just inside the incident and exit surfaces of the wave plate. The phase changes of the o and e waves after traversing the thickness of the wave plate are:

$$\begin{bmatrix} E'_o \\ E'_e \end{bmatrix} = \begin{bmatrix} E_o e^{-ik_o\ell} \\ E_e e^{-ik_e\ell} \end{bmatrix}, \tag{7.83}$$

where the identical phase factor $e^{i(\omega t - kz)}$ for the two waves at the input side is omitted at the right hand side of Eq. (7.83). Equation (7.83) can be rewritten as:

$$\begin{bmatrix} E'_o \\ E'_e \end{bmatrix} = \begin{bmatrix} e^{-ik_o\ell} & 0 \\ 0 & e^{-ik_e\ell} \end{bmatrix} \begin{bmatrix} E_o \\ E_e \end{bmatrix}, \tag{7.84}$$

Let

$$M_p = \begin{bmatrix} e^{-ik_o\ell} & 0 \\ 0 & e^{-ik_e\ell} \end{bmatrix}. \tag{7.85}$$

It can be seen that M_p represents a phase shift operator (matrix). Summarizing the above, we can write down the total operation W representing the propagation through a wave plate as:

$$W = R(-\psi)M_pR(\psi), \tag{7.86}$$

and we have

$$\begin{bmatrix} E'_x \\ E'_y \end{bmatrix} = W \begin{bmatrix} E_x \\ E_y \end{bmatrix}, \tag{7.87}$$

where the two Jones vectors on the left and right sides of Eq. (7.87) are the field components of the transmitted and incident waves in the laboratory frame, respectively. The operator W of Eq. (7.87) means that the incident vector is first operated by $R(\psi)$, then by M_p, and then by $R(-\psi)$. It should be stressed that the order of such operations should start from the right to the left. We can simplify M_p slightly by defining:

$$\epsilon \equiv (k_o - k_e)l \equiv \text{relative phase shift} \tag{7.88}$$

and

$$\phi = \frac{1}{2}(k_o + k_e)\ell \equiv \text{mean absolute phase change.} \tag{7.89}$$

Thus:

$$\frac{\epsilon}{2} + \phi = k_o\ell \tag{7.90}$$

and

$$-\frac{\epsilon}{2} + \phi = k_e\ell \tag{7.91}$$

and from Eq. (7.85):

$$M_p = e^{-i\phi} \begin{bmatrix} e^{-i\epsilon/2} & 0 \\ 0 & e^{i\epsilon/2} \end{bmatrix}. \tag{7.92}$$

Thus:

$$W = e^{-i\phi} \begin{bmatrix} \cos\psi & -\sin\psi \\ \sin\psi & \cos\psi \end{bmatrix} \begin{bmatrix} e^{-i\epsilon/2} & 0 \\ 0 & e^{i\epsilon/2} \end{bmatrix} \begin{bmatrix} \cos\psi & \sin\psi \\ -\sin\psi & \cos\psi \end{bmatrix} \tag{7.93'}$$

and

$$\begin{bmatrix} E'_x \\ E'_y \end{bmatrix} = W \begin{bmatrix} E_x \\ E_y \end{bmatrix}. \tag{7.93}$$

It is left an <u>exercise</u> for the reader to show that W is a unitary matrix, i.e., $w^t w = I$. Thus, if (E_x, E_y) is normalized, (E'_x, E'_y) is also a normalized vector.

For a *dichroic polarizer* (by absorption), such as Polaroid sheets, the matrix representation for the propagation of an EM wave is something else. Let A be the absorbing (non-birefringent) polarizer's operator (matrix). One sees that

$$\begin{bmatrix} E'_{N_x} \\ E'_{N_y} \end{bmatrix} = A \begin{bmatrix} E_{N_x} \\ E_{N_y} \end{bmatrix}. \tag{7.94}$$

Since this type of polarizer gives only linear polarization, i.e., either $E'_x = 0$ or $E'_y = 0$, there are only two expressions for A, namely:

$$A_x = e^{-i\phi} \begin{bmatrix} 1 & 0 \\ 0 & 0 \end{bmatrix}, \tag{7.95}$$

$$A_y = e^{-i\phi} \begin{bmatrix} 0 & 0 \\ 0 & 1 \end{bmatrix}, \tag{7.96}$$

where $e^{-i\phi}$ is a general phase factor, and A_x and A_y are the matrices that will transform the incident wave into a wave of linear polarization along the x and y-axes respectively. To show this, one simply substitutes Eqs. (7.95) and (7.96) into (7.94) and obtains:

$$\begin{bmatrix} E'_{N_x} \\ E'_{N_y} \end{bmatrix} = \begin{cases} A_x \begin{bmatrix} E_{N_x} \\ E_{N_y} \end{bmatrix} \\ A_y \begin{bmatrix} E_{N_x} \\ E_{N_y} \end{bmatrix} \end{cases} = \begin{cases} e^{-i\phi} E_{N_x} \begin{bmatrix} 1 \\ 0 \end{bmatrix} \\ e^{-i\phi} E_{N_y} \begin{bmatrix} 0 \\ 1 \end{bmatrix} \end{cases}.$$

Example 1: *Quarter-wave plate*

Similar to Section 7.7, the plate introduces a relative phase shift of

$$\epsilon = 2n\pi \pm \pi/2.$$

Using Eq. (7.93), assuming that the incident wave is linearly polarized in the y-direction, and $\psi = 45°$, we have:

$$\begin{bmatrix} E'_x \\ E'_y \end{bmatrix} = W \begin{bmatrix} 0 \\ 1 \end{bmatrix},$$

$$= \cdots .(\underline{\text{exercise}}\text{ for the reader})$$

$$= e^{-i\phi} \begin{bmatrix} \cos(\pm\pi/4) & -i\sin(\pm\pi/4) \\ -i\sin(\pm\pi/4) & \cos(\pm\pi/4) \end{bmatrix} \begin{bmatrix} 0 \\ 1 \end{bmatrix}, \qquad (7.97)$$

$$= e^{-i\phi} \begin{bmatrix} -i\sin(\pm\pi/4) \\ \cos(\pm\pi/4) \end{bmatrix},$$

$$= e^{-i\phi} \begin{bmatrix} \mp i\frac{1}{\sqrt{2}} \\ \frac{1}{\sqrt{2}} \end{bmatrix},$$

$$= \frac{e^{-i\phi}}{\sqrt{2}} \begin{bmatrix} \mp i \\ 1 \end{bmatrix}.$$

Using the x-component as the reference, we have:

$$\begin{bmatrix} E'_x \\ E'_y \end{bmatrix} = \frac{ie^{-i\phi}}{\sqrt{2}} \begin{bmatrix} 1 \\ \pm i \end{bmatrix} \rightarrow \frac{1}{\sqrt{2}} \begin{bmatrix} 1 \\ \pm i \end{bmatrix}, \qquad (7.98)$$

where the last step omits the common phase for normalization purposes. In other words, a quarter-wave plate transforms a linear polarization to circular polarization if $\psi = 45°$. Since the plate is passive, the reverse should also be true, i.e., it turns circular polarization into a linear one. To prove this, we observe from Eq. (7.97) that

$$W = e^{-i\phi} \begin{bmatrix} \cos(\pm\pi/4) & -i\sin(\pm\pi/4) \\ -i\sin(\pm\pi/4) & \cos(\pm\pi/4) \end{bmatrix}, \qquad (7.99)$$

$$= \frac{e^{-i\phi}}{\sqrt{2}} \begin{bmatrix} 1 & \mp i \\ \mp i & 1 \end{bmatrix}, \tag{7.100}$$

$$\begin{bmatrix} E'_x \\ E'_y \end{bmatrix} = W \cdot \frac{1}{\sqrt{2}} \begin{bmatrix} 1 \\ \pm i \end{bmatrix} \qquad \text{(by Eq. (7.74))}$$

$$= \frac{e^{-i\phi}}{2} \begin{bmatrix} 1 & \mp i \\ \mp i & 1 \end{bmatrix} \begin{bmatrix} 1 \\ \pm i \end{bmatrix}. \tag{7.101}$$

Equation (7.101) gives:

$$\begin{bmatrix} E'_x \\ E'_y \end{bmatrix} = \left\{ \begin{array}{l} \frac{e^{-i\phi}}{2} \begin{bmatrix} 2 \\ 0 \end{bmatrix} = e^{-i\phi} \begin{bmatrix} 1 \\ 0 \end{bmatrix} \\[3mm] \frac{e^{i\phi}}{2} \begin{bmatrix} 0 \\ \pm 2i \end{bmatrix} = \pm i e^{-i\phi} \begin{bmatrix} 0 \\ 1 \end{bmatrix} \end{array} \right\}. \tag{7.102}$$

Omitting the common phases in Eq. (7.102) for normalization purposes, we have:

$$\begin{bmatrix} E'_x \\ E'_y \end{bmatrix} = \left\{ \begin{array}{l} \begin{bmatrix} 1 \\ 0 \end{bmatrix} \\[3mm] \begin{bmatrix} 0 \\ 1 \end{bmatrix} \end{array} \right. = \left\{ \begin{array}{l} J_{Nn} \\ J_{NV} \end{array} \right., \tag{7.103}$$

which proves the above statement.

Example 2: *Half-wave plate*

As in Section 7.7, the plate introduces a relative phase shift of:

$$\epsilon = (2n + 1)\pi, (n = 0, 1, 2, \ldots). \tag{7.104}$$

For an arbitrary angle ψ between the x and the o-axes, Eq. (7.93) becomes, starting with a vertically polarized incident wave,

$$\text{i.e.,} \quad \begin{bmatrix} E_x \\ E_y \end{bmatrix} = \begin{bmatrix} 0 \\ 1 \end{bmatrix},$$

$$\begin{bmatrix} E'_x \\ E'_y \end{bmatrix} = i e^{-i\phi} \begin{bmatrix} \cos(\pi/2 + 2\psi) \\ \sin(\pi/2 + 2\psi) \end{bmatrix}. \tag{7.105}$$

The derivation is left as an <u>exercise</u> to the reader. Omitting the common phase factor for normalization purposes, we obtain:

$$\begin{bmatrix} E'_x \\ E'_y \end{bmatrix} = \begin{bmatrix} \cos(\pi/2 + 2\psi) \\ \sin(\pi/2 + 2\psi) \end{bmatrix}. \qquad (7.106)$$

This is a linear polarization making an angle $(\pi/2 + 2\psi)$ with the x-axis (see Eq. (7.66)). Since the incident wave's polarization makes an angle of $\pi/2$ with the x-axis, the transmitted polarization (being still linear) makes an angle of 2ψ with respect to the incident polarization. That is, *a half-wave plate rotates a linear polarization by an angle of* $+2\psi$. If $\psi = 45°$, the angle of rotation is $90°$, which is the case commonly used in experiments. Note that the direction of rotation of the polarization is the same as that of ψ (the rotation of the x-axis into the o-axis, see Fig. 7.12).

Some comments on ψ

The angle ψ between the x- and o-axes depends on the orientation of the optic axis. Since, as defined in Section 7.7, the optic axis of a uniaxial wave plate is perpendicular to z, and since the polarization of the o-wave is always perpendicular to the optic axis and the propagation direction (Fig. 7.13(a)) (see also the section on the index ellipsoid in Chapter 6), the polarization of the e-wave is in the plane containing the optic axis. In practice, the wave plate is rotated about the z-axis so that the o-axis makes an angle of ψ with the laboratory x-axis, (Fig. 7.13(b)). This angle is usually set at $45°$ so that a quarter-wave plate will transform linear polarization into circular, and vice versa, and a half-wave plate will rotate a linear polarization by $90°$.

The angle $\psi = 45°$ can be experimentally determined in the following way. Since ψ is the angle between the o-wave's polarization and the x-axis (horizontal, for instance), we need to determine the o-axis. The o-axis is the direction of polarization of the o-wave. By definition, it is always perpendicular to the optic axis. Hence, we should determine the optic axis. Figure 7.14 shows one way of determining the optic axis. We start with a pair of crossed polarizers (Fig. 7.14(a)). A light beam of arbitrary polarization is sent into the crossed polarizers with their polarization transmission axes v and n as indicated. No light will pass through the second polarizer. Now,

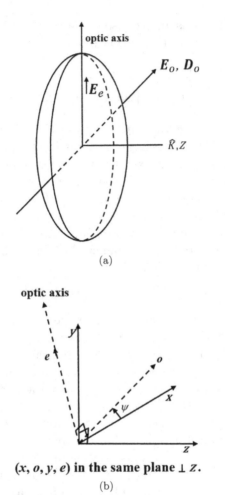

(a)

(x, o, y, e) in the same plane ⊥ z.

(b)

Fig. 7.13.   Analysis of the angle ψ between the o-wave and the laboratory x-axis of a uniaxial wave plate.

we add our wave plate in between the crossed polarizer. An arbitrarily oriented wave plate will generate two waves (o and e) with orthogonal polarizations so that at the exit of the wave plate W, the polarization state is changed with respect to the input polarization, which is perpendicular to the paper. Thus, one can decompose the polarization at the output of W into two polarizations preferred by P_1, and P_2, i.e., v and n directions. At the exit of P_2, we detect some light polarized in the n-direction.

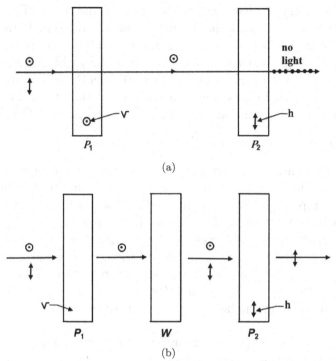

Fig. 7.14. Experimental determination of the angle ψ between the o-wave and the x-axis of the laboratory of a uniaxial wave plate.

We now rotate W so that at the output of P_2, the transmitted intensity becomes zero again. This situation corresponds to two possibilities in W — either the optic axis is parallel to the propagation direction, or it is perpendicular to the propagation direction. When the optic axis is parallel to the direction of propagation, by definition, the polarization state of the light is unchanged in passing through W, as if W is not there. Hence, no light will get through P_2. If the optic axis is perpendicular to the propagation direction and, in particular, in the polarization transmission direction n of P_2, the incident V-polarization will pass through W because it is perpendicular to the optic axis and thus represents the o-wave. Since there is no e-component in the incident beam to W, only V-polarization comes out from W, and P_2 will absorb or reject it. Similarly, if the optic axis is in the direction V of P_1, the incident V-polarization will pass through W because it is now an e-wave. P_2 again absorbs it.

We now keep W fixed at such a position and rotate P_1 only by an angle $\theta \neq 90°$ about the z-axis (propagation axis). If the optic axis in W is parallel to the propagation direction, any polarization will pass through W without any change, and P_2 will stop it, i.e., no transmission after P_2. If the optic axis in W is in the n or V-direction, there will be o and e waves generated in W so that P_2 will transmit the n-polarization component. Thus, the optic axis is determined.

Some comments on slow and fast axes

At the beginning of this section, we very briefly mentioned fast and slow axes. They can be confusing if one is not careful. The fast and slow axes represent the polarization directions of the o and e waves inside a uniaxial crystal. If the o-wave propagates faster than the e-wave (positive uniaxial crystal), the polarization vector of the o-wave is along the fast axis, and that of the e-wave is along the slow axis. For a negative uniaxial crystal, the reverse is true. That is, since the e-wave propagates faster than the o-wave, the e-wave's polarization vector is along the fast axis, and that of the o-wave is along the slow axis.

In all, the fast (slow) axis represents the faster (slower) polarization direction of the o and e waves.

7.10 The Power of Crossed Polarizers

A pair of crossed polarizers are simply two polarizers of any type whose transmission axes are perpendicular to each other so that the light of some appropriate wavelengths and of any polarization will not pass through them. (The light will pass through one but will be blocked by the other.) It becomes essentially a light isolator (Fig. 7.15(a)). If one inserts between the two polarizers an optical element that alters the state of polarization of the light that is inside there, the light whose polarization is altered will be transmitted. Such an idea has been put to good use in many optical systems. The following are some examples.

Figure 7.15(b) shows a pair of crossed polarizers with an electro-optical element in between. When appropriately biased, the electro-optical (EO) crystal can act as a quarter-wave plate, for example, and the output of the ensemble follows the voltage change across the

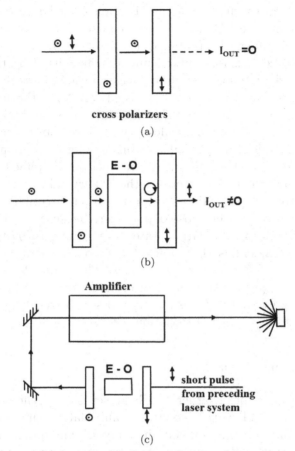

Fig. 7.15. (a) and (b) An example of the application of a cross polarizer with an electro-optic element. (c) A cross polarizer together with an electro-optic element acts as a back reflection isolator in a laser oscillator-amplifiers chain.

EO crystal, i.e., modulation. The crossed polarizer can be applied to strain analysis in matter (Section 9.2, last paragraph), Q-switching, mode locking, pulse slicing, pulse selection, back reflection isolator in laser amplifier chains, etc. Some of the above items will be explained in Chapter 8.

We discuss the last item, i.e., back reflection isolation. This is shown in Fig. 7.15(c). When the laser pulse (short, less than a few tens of nanoseconds) arrives at the device, the EO element is quickly biased to V_π so that the laser pulse, polarized vertically, after being transmitted by the first polarizer, is transformed into horizontally polarized light (the EO element is of course properly oriented with its optic axis along the propagation direction. It acts as an electrically induced half-wave plate) and is transmitted by the second crossed polarizer. Soon after the passage of the laser pulse, the high voltage across the EO crystal is turned off. Now, the laser pulse, after passing through the amplifier, hits the target, and there is some reflection that will come back along the original optical path. If the amplifier's gain is still significant at this moment, it will amplify the back-scattered pulse significantly. But such an amplified pulse will not go through the isolator because the EO crystal is now not biased and is isotropic (optic axis along propagation direction). Any light will not go through the crossed polarizer, because the EO crystal does not affect the polarization state of the light passing through it.

A final comment is the application of polarized light to create the sensation of *3D pictures or movies*. The principle is based on two sets of images emitting light of crossed linearly polarized light. Hence, using eyeglasses transmitting cross-polarized light into each eye would result in the 3D sensation.

7.11 Closing Remarks

This chapter presents a rather detailed discussion on the polarization states of light and how they could be manipulated using wave plates. The Jones matrix formation helps in analyzing and designing systems with both active and passive polarizing components.

Reference

E. Hecht (2017). *Optics*, Fifth edition, Addison-Wesley.

Chapter 8

Electric Field-Induced Anisotropy: Electro-Optics and Q-switching

An external electric field will induce in a dielectric a polarization, which in general is anisotropic, whether or not the material is naturally anisotropic. Depending on the electric field strength and the material structure, some anisotropy is negligible, but some is significant. When light (laser) waves propagate through a transparent dielectric, to which an external electric field is applied across, the state of polarization of the light waves will be changed because of the natural and induced anisotropy of the material. Because one can vary the external electric field at rather high frequencies, the variation of the state of polarization of light waves will thus be fast. This means that one, through the manipulation of the state of polarization or the phase of the light, can electrically manipulate light waves at a fast rate. If one can find a detector to detect such changes, such manipulation or modulation can be applied to high-speed signal processing, communication, short laser pulse generation, fast shutter, etc. High-speed modulation is possible in principle because the frequency of the optical wave (carrier wave) is very high ($\sim 10^{14}$ Hz). Thus, coupled with laser beams, such so-called electro-optical (EO) devices become one of the principal elements in a photonic or optoelectronic system.

8.1 Electric Field-Induced Anisotropy

A DC electric field is applied across an anisotropic material. Normally, the external electric field is weak compared to the internal fields of the material (usually crystal). Hence, we can consider the external electric field as a perturbation to the internal field in later calculations.

The external DC field induces a polarization in the anisotropic medium. If we now send in an EM wave, it will "detect" that the original unperturbed dielectric tensor is modified (by the DC field-induced polarization). This means that the original index ellipsoid describing the response of the unperturbed medium to the EM wave will be modified. In this chapter, we shall study how the index ellipsoid is changed and how this is used in applications. Essentially, the ellipsoid will be deformed and tilted through a rotation in space with respect to the original ellipsoid. We now carefully analyze what happens.

Originally, the index ellipsoid in the principal coordinate axes is given by (see Chapter 6):

$$\frac{x^2}{n_x^2} + \frac{y^2}{n_y^2} + \frac{z^2}{n_z^2} = 1, \tag{8.1}$$

where

$$n_i^2 \equiv \frac{\epsilon_i}{\epsilon_0} \quad (i = x, y, z), \tag{8.2}$$

$$(x, y, z) \equiv (D_X, D_Y, D_z) \frac{1}{\sqrt{\epsilon_0 2 w_e}}. \tag{8.3}$$

The external DC electric field induces a change in the polarization, dielectric tensor, indices, and field D. The new dielectric tensor will no longer be diagonal, and the equation of the new index ellipsoid will contain cross-terms in the original principal coordinate axes. It is more convenient to use the notion of impermeability to describe the index ellipsoid (Eq. (8.1)).

Definition: *The impermeability tensor*

$$\eta_{ij} \equiv \epsilon_0 / \epsilon_{ij}, \tag{8.4}$$

where $\epsilon_{ij} = \epsilon_x, \epsilon_y \epsilon_z$ for $i = x, y, z$, respectively. Substituting Eq. (8.4) into (8.1), one obtains:

$$\eta_{xx} x^2 + \eta_{yy} y^2 + \eta_{zz} z^2 = 1. \tag{8.5}$$

With the external electric field applied to the medium, Eq. (8.5) now changes into one containing cross terms,

$$\text{i.e.,} \quad \eta_{ij} x_i x_j = 1, \tag{8.6}$$

where η_{ij} is a function of E and repeated indices represent summation over the indices. For simplicity, the coordinates x, y and z are replaced by x_i or x_j where $i,j = 1, 2, 3$. In what follows, (x, y, z) and (x_1, x_2, x_3) will be interchanged freely, using whichever is more convenient.

η_{ij} is now a function of the external field E, i.e., $\eta_{ij}(E)$. Because we have assumed that the external field induces only a perturbation in the permeability, we can express $\eta_{ij}(E)$ in a Taylor series around the field free value (i.e., at $E = o$).

$$\eta_{ij}(E) = \eta_{ij}(0) + \sum_k \left(\frac{\partial \eta_{ij}}{\partial E_k} \right)_{E_k=0} (E_k - 0)$$

$$+ \frac{1}{2} \sum_{k,l} \left(\frac{\partial^2 \eta_{ij}}{\partial E_k \partial E_l} \right)_{E_k=E_l=0} (E_k - 0)(E_\ell - 0)$$

$$+ \cdots \tag{8.7}$$

Rewriting Eq. (8.7) as

$$\eta_{ij}(E) - \eta_{ij}(0) \equiv \Delta \eta_{ij}(E)$$

$$= \sum_k (r_{ij})_k E_k \tag{8.8}$$

$$+ \sum_{k,\ell} (s_{ij})_{kl} E_k E_\ell + \cdots$$

where

$$(r_{ij})_k \equiv \left(\frac{\partial \eta_{ij}}{\partial E_k} \right)_{E_k=0,} \quad (s_{ij})_{k\ell} \equiv \frac{1}{2} \left(\frac{\partial^2 \eta_{ij}}{\partial E_k \partial E_l} \right)_{E_k=E_\ell=0} . \tag{8.9}$$

Equation (8.8) represents the electro-optic response of the medium. Using repeated indices to represent summation, Eq. (8.8) becomes

$$\Delta \eta_{ij}(E) = r_{ijk} E_k + s_{ijk\ell} E_k E_\ell + \cdots , \tag{8.10}$$

where $r_{ijk} \equiv (r_{ij})_k \equiv$ *linear or Pockels electro-optic coefficient* and $s_{ijk\ell} \equiv (s_{ij})_{k\ell} \equiv$ *quadratic or Kerr electro-optic coefficient*.

We note that when $E = 0$, Eq. (8.6) becomes

$$\eta_{ij}(0)x_i x_j = 1, \tag{8.11}$$

which should be identical to Eq. (8.5), i.e.,

$$\eta_{ij}(0)x_i x_j = \eta_{\mathrm{xx}}x^2 + \eta_{\mathrm{yy}}y^2 + \eta_{\mathrm{zz}}z^2,$$

$$= \frac{x^2}{n_{\mathrm{x}}^2} + \frac{y^2}{n_{\mathrm{y}}^2} + \frac{z^2}{n_{\mathrm{z}}^2}. \tag{8.12}$$

8.2   Linear EO Effect: Pockels Effect

In the expression for the electro-optic response of the medium (Eq. (8.8) or (8.10)), if the first term at the right hand side is dominant, then we can neglect the higher-order perturbation terms and obtain

$$\Delta\eta_{ij}(E) \equiv r_{ijk}E_k. \tag{8.13}$$

This is the *linear electro-optic response* (*Pockels effect*). We note that since ϵ_{ij} is a symmetric tensor, η_{ij} (Eq. (8.4)) must also be symmetric, i.e.,

$$\eta_{ij} = \eta_{ji}(i \neq j). \tag{8.14}$$

Hence,

$$r_{ijk} \equiv \left(\frac{\partial\eta_{ij}}{\partial E_k}\right)_{E_k=0}, \quad \text{(by Eq. (8.9))},$$

$$= \left(\frac{\partial\eta_{ji}}{\partial E_k}\right)_{E_k=0}, \quad \text{(by Eq.(8.14))},$$

$$= r_{jik}. \tag{8.15}$$

We can now write the equation of the modified index ellipsoid (Eq. (8.6)) as follows:

$$\eta_{ij}(E)x_i x_j = 1,$$

$$[\eta_{ij}(0) + \Delta\eta_{ij}(E)]x_i x_j = 1,$$

$$\frac{x^2}{n_{\mathrm{x}}^2} + \frac{y^2}{n_{\mathrm{y}}^2} + \frac{z^2}{n_{\mathrm{z}}^2} + r_{ijk}E_k x_i x_j = 1. \tag{8.16}$$

$$r_{ijk}E_k x_i x_j \equiv \sum_{i,j,k=1}^{3} r_{ijk}E_k x_i x_j,$$

$$\equiv \sum_{i,j} x_i x_j \left(\sum_k r_{ijk}E_k \right),$$

$$= \sum_{i,j} x_i x_j \left(r_{ij1}E_1 + r_{ij2}E_2 + r_{ij3}E_3 \right),$$

$$= \cdots (\underline{\text{exercise}}).$$

$$= (x_1,x_2,x_3) \begin{pmatrix} (x_1,x_2,x_3) \begin{pmatrix} r_{111} & r_{112} & r_{113} \\ r_{121} & r_{122} & r_{123} \\ r_{131} & r_{132} & r_{133} \end{pmatrix} \\ (x_1,x_2,x_3) \begin{pmatrix} r_{211} & r_{212} & r_{213} \\ r_{221} & r_{222} & r_{223} \\ r_{231} & r_{232} & r_{233} \end{pmatrix} \\ (x_1,x_2,x_3) \begin{pmatrix} r_{311} & r_{312} & r_{313} \\ r_{321} & r_{322} & r_{323} \\ r_{331} & r_{332} & r_{333} \end{pmatrix} \end{pmatrix} \begin{pmatrix} E_1 \\ E_2 \\ E_3 \end{pmatrix},$$

$$(8.17)$$

where Eqs. (8.12) and (8.13) were used. We expand the fourth term on the left hand side of Eq. (8.16).

The matrix multiplication of Eq. (8.17) should be done from the right to left. We use the following short hand notation to simplify the triple subscripts of r.

(ij)	Short Hand Notation
(11)	1
(22)	2
(33)	3
(23) = (32)	4
(13) = (31)	5
(12) = (21)	6

The last three rows are permitted because of the symmetry of the permeability tensor (Eqs. (8.14) and (8.15)). With these, Eq. (8.17) can be simplified. (This is an <u>exercise</u> for the reader.)

$$M \equiv r_{ijk}E_k x_i x_j,$$

$$\equiv (x_1^2, \ x_2^2, \ x_3^2, \ 2x_2x_3, \ 2x_1x_3, \ 2x_1x_2) \begin{pmatrix} r_{11} & r_{11} & r_{13} \\ r_{21} & r_{22} & r_{23} \\ r_{31} & r_{32} & r_{33} \\ r_{41} & r_{42} & r_{43} \\ r_{51} & r_{52} & r_{53} \\ r_{61} & r_{62} & r_{63} \end{pmatrix} \begin{pmatrix} E_1 \\ E_2 \\ E_3 \end{pmatrix}.$$

$$(8.18)$$

(The matrix r_{ij} is called the electro-optic tensor or coefficient.) Hence, the general index ellipsoid under the linear electro-optic effect is, from Eq. (8.16), using $(x, y, z) = (x_1, x_2, x_3)$, for the sake of uniformity:

$$\frac{x_1^2}{n_{x_1^2}} + \frac{x_2^2}{n_{x_2^2}} + \frac{x_3^2}{n_{x_3^2}} + M = 1, \tag{8.19}$$

where the 4th term, M, on the left hand side, is given by Eq. (8.18). Normally, the matrix r_{ij} in Eq. (8.18) is known for an EO material.

Example: *A KDP crystal*
    KDP $\equiv$ Potassium dihydrogen phosphate or KH_2PO_4.
    This is a very popular crystal and its r_{ij} matrix is known:

$$r_{ij} = \begin{pmatrix} 0 & 0 & 0 \\ 0 & 0 & 0 \\ 0 & 0 & 0 \\ r_{41} & 0 & 0 \\ 0 & r_{41} & 0 \\ 0 & 0 & r_{63} \end{pmatrix}. \tag{8.20}$$

Values of the elements of r_{ij} for various practical crystals have been tabulated (see Yariv and Yeh, 1984).

Substituting into Eq. (8.18) gives:

$$r_{ijk}E_k x_i x_j = 2r_{41}E_1 x_2 x_3 + 2r_{41}E_2 x_1 x_3 + 2r_{63}E_3 x_1 x_2. \qquad (8.21)$$

Substituting into Eq. (8.19), and changing (x_1, x_2, x_3) back to (x, y, z) for the sake of convenience, we have

$$\frac{x^2}{n_x^2} + \frac{y^2}{n_y^2} + \frac{z^2}{n_z^2} + 2r_{41}E_1 yz + 2r_{41}E_2 xz + 2r_{63}E_3 xy = 1. \qquad (8.22)$$

KDP is a uniaxial crystal

$$\therefore n_x = n_y = n_o, n_z \equiv n_e. \qquad (8.23)$$

Equation (8.22) becomes:

$$\frac{x^2}{n_o^2} + \frac{y^2}{n_o^2} + \frac{z^2}{n_e^2} + 2r_{41}E_1 yz + 2r_{41}E_2 xz + 2r_{63}E_3 xy = 1. \qquad (8.24)$$

Case 1: Electric field in the z-direction
    i.e.,

$$E_1 = E_2 = 0, E_3 = E_z \neq 0. \qquad (8.25)$$

See Fig. 8.1(a). Equation (8.24) becomes:

$$\frac{x^2}{n_o^2} + \frac{y^2}{n_o^2} + \frac{z^2}{n_e^2} + 2r_{63}E_z xy = 1. \qquad (8.26)$$

This is the equation of an ellipsoid. Rotation of the axes (see below) will bring Eq. (8.26) into the following familiar form:

$$\frac{x'^2}{n_{x'}^2} + \frac{y'^2}{n_{y'}^2} + \frac{z'^2}{n_{z'}^2} = 1, \qquad (8.27)$$

where (x', y', z') is the new principal coordinate axes with $n_{x'}$, $n_{y'}$ and $n_{z'}$ the corresponding indices. To derive Eq. (8.27), we need to find the analytic expressions for n'_x, etc. We make the following observations:

(a) In both Eqs. (8.26) and (8.27), the xz and yz terms are missing. The rotation does not involve any rotation of the z-axis; hence, $z = z'$. Thus, $n_e^2 = n_{z'}^2$ (see Fig. 8.1(b) and (c)).

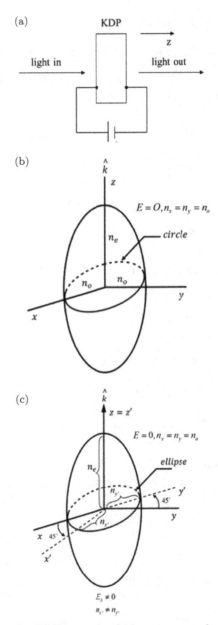

Fig. 8.1.   Analysis of a KDP crystal, which acts as an electro-optic wave plate (Pockels effect). In (a), the z-axis is also the optic axis.

(b) The rotation is thus a rotation of the x, y axes about the z-axis (from observation (a)). Let the angle of rotation be α. From analytic geometry, every quadratic expression

$$Ax^2 + Bxy + Cy^2, (B \neq 0), \tag{8.28}$$

can be reduced to the form:

$$A'x'^2 + C'y'^2$$

by rotating the axes through an angle α $(0 < \alpha < 90°)$, where

$$\cot 2\alpha = \frac{A - C}{B}. \tag{8.29}$$

From Eqs. (8.26) and (8.28), we see that $A = C$ in Eq. (8.29). Hence,

$$\cot 2\alpha = 0,$$

$$\alpha = 45° \tag{8.30}$$

and

$$x = x' \cos \alpha - y' \sin \alpha, \tag{8.31}$$

$$y = x' \sin \alpha + y' \cos \alpha. \tag{8.32}$$

Substituting Eqs. (8.31) and (8.32) with (8.30) into (8.26) gives (exercise):

$$\left(\frac{1}{n_0^2} + r_{63} E_z \right) x'^2 + \left(\frac{1}{n_0^2} - r_{63} E_z \right) y'^2 + \frac{z'^2}{n_e^2} = 1. \tag{8.33}$$

Alternative way of deriving Eq. (8.33):

This is a general way of diagonalizing the symmetric matrix

$$S = \begin{pmatrix} a & d & e \\ d & b & f \\ e & f & c \end{pmatrix} \tag{8.34}$$

of the general quadratic equation:

$$ax^2 + by^2 + cz^2 + 2dxy + 2exz + 2fyz = g. \tag{8.35}$$

The diagonalization requires that

$$|S - \lambda I| = 0, \tag{8.36}$$

where I is a unitary matrix and λ is a root of the diagonal matrix of S, i.e.,

$$S_D \equiv \begin{bmatrix} \lambda_1 & 0 & 0 \\ 0 & \lambda_2 & 0 \\ 0 & 0 & \lambda_3 \end{bmatrix}. \tag{8.37}$$

The reader can solve Eq. (8.36) as an <u>exercise</u> using the definitions in Eqs. (8.34) and (8.35) and comparing with Eq. (8.26).

Comparing Eqs. (8.33) and (8.27), we have:

$$\frac{1}{n_{x'}^2} = \frac{1}{n_0^2} + r_{63}E_z, \tag{8.38}$$

$$\frac{1}{n_{y'}^2} = \frac{1}{n_0^2} - r_{63}E_z, \tag{8.39}$$

$$n_{z'} = n_e. \tag{8.40}$$

Since the change in the index by the electric field is assumed to be a perturbation (i.e., small), we can write:

$$\frac{1}{n_i^2} - \frac{1}{n_0^2} = \Delta\left(n^{\frac{1}{2}}\right)\bigg|_{n=n_o}, (\Delta : \text{differential}, i = x' \text{ or } y'),$$

$$= -2n^{-3}|_{n=n_o}\Delta n, \tag{8.41}$$

$$= -2n_o^{-3}(n_i - n_o),$$

$$n_i - n_o = -\frac{1}{2}n_o^3\left(\frac{1}{n_i^2} - \frac{1}{n_0^2}\right). \tag{8.42}$$

Substituting Eqs. (8.38) and (8.39) into (8.42) separately, one obtains:

$$n_{x'} = n_o - \frac{1}{2}n_o^3 r_{63}E_z, \tag{8.43}$$

$$n_{y'} = n_o + \frac{1}{2}n_o^3 r_{63}E_z, \tag{8.44}$$

$$n_{z'} = n_e. \tag{8.45}$$

Case 2: Electric field in the x-direction i.e.,

$$E_1 \equiv E_x \neq 0, E_2 = E_3 = 0. \tag{8.46'}$$

Substituting into Eq. (8.24), we have:

$$\frac{x^2}{n_o^2} + \frac{y^2}{n_o^2} + \frac{z^2}{n_e^2} + 2r_{41}E_x yz = 1. \tag{8.46}$$

Using a similar analysis as in Case 1, one can rotate the y, z axes about the x-axis and transform Eq. (8.46) into the "normal" form of an ellipsoid.

The detail is left as an <u>exercise</u>. Letting α be the angle through which the y, z axes are rotated, one obtains

$$\tan 2\alpha = \frac{2r_{41}E_x}{\frac{1}{n_{xo}^2} - \frac{1}{n_e^2}} \tag{8.47}$$

and the ellipsoid is

$$\frac{x'^2}{n_o^2} + y'^2 \left\{ \frac{1}{n_o^2} + \frac{\tan^2 \alpha}{n_e^2} + 2r_{41}E_x \tan \alpha \right\} \cos^2 \alpha$$

$$+ z'^2 \left\{ \frac{1}{n_e^2} + \frac{\tan^2 \alpha}{n_o^2} - 2r_{41}E_x \tan \alpha \right\} \cos^2 \alpha = 1. \tag{8.48}$$

From tabulated values (Yariv and Yeh, 1984):

$$r_{41} \approx 9 \times 10^{-12}\,\text{m/V} \quad \text{at } \lambda \cong 0.55\,\mu\text{m}.$$

Hence, even at a very high DC voltage, e.g., $E_x = 10^6\,\text{V/m}$, $r_{41}E_x \approx 10^{-5}$, a very small value, while normally, $1/n_o^2 - 1/n_e^2 \sim 10^{-1}$.

Hence,

$$\tan 2\alpha \approx 0, \tag{8.49}$$

$$\cos^2 \alpha \approx 1, \tag{8.50}$$

$$\tan^2 \alpha \approx 0. \tag{8.51}$$

Using Eqs. (8.50) and (8.51) in (8.48) yields

$$\frac{x'^2}{n_o^2} + y'^2 \left\{ \frac{1}{n_o^2} + 2r_{41}E_x \tan \alpha \right\} + z'^2 \left\{ \frac{1}{n_e^2} - 2r_{41}E_x \tan \alpha \right\} = 1. \tag{8.52}$$

Again, using the same procedure as in Eqs. (8.41) and (8.42), the new indices are obtained (underline):

$$n_{x'} = n_o, \tag{8.53}$$

$$n_{y'} = n_o - \frac{1}{2}n_o^3 r_{41} E_x \tan \alpha, \tag{8.54}$$

$$n_{z'} = n_e + \frac{1}{2}n_e^3 r_{41} E_x \tan \alpha. \tag{8.55}$$

8.3   Application to Electrical Modulation of Light Waves: EO Modulator

Consider a z-cut[1] KDP crystal plate (the propagation direction of the electromagnetic (EM) wave is also along the z-axis). We apply an AC electric field across it.

$$\boldsymbol{E}(t) = E_z(t)\hat{z}. \tag{8.56}$$

The index ellipsoid of the crystal will thus change in time as the electric field varies. In this book, we study only cases in which the frequency of the electric field is not too high, so that the index change faithfully "follows" the variation of the external electric field, i.e., the index change and the electric field are in phase.

Case 1: DC field

Assume for the moment that the external field is still DC. From Section 8.2, the modified index ellipsoid of the KDP crystal under a DC electric field directed along the z-direction is given by Eq. (8.33), and Eqs. (8.43) to (8.45) give the modified indices (see also Eq. (8.27)). Since the propagation direction $\hat{k}$ is along the principal z-axis and $z = z'$, we obtain the intersection ellipse between the x', y' plane perpendicular to $\hat{k}$ and the modified index ellipsoid shown in Fig. 8.1(c). Note that the modified ellipsoid does not possess a rotational symmetry about the z'-axis because $n_{x'} = n_{y'}$. Thus, the EM wave propagating in the z-direction sees two indices $n_{x'}$ and $n_{y'}$, i.e., there are

---

[1]z-cut: The principal z-axis of the index ellipsoid of the crystal is perpendicular to the crystal (plate) surface. In this case of KDP, which is a uniaxial crystal, the z-axis is also the optic axis (also called the c-axis), around which the index ellipsoid has a rotational symmetry (see Chapter 6).

two waves propagating at two different velocities c/n'_x and $c/n_{y'}$ in the crystal. Since the incident direction is perpendicular to the crystal plate surface, the two waves in the crystal propagate in the same direction (z-axis). The relative phase retardation (shift) between the two waves is (see Section 7.7, Eq. (7.32)):

$$\epsilon = (k'_y - k'_x)d, \tag{8.57}$$

where d is the thickness of the plate,

$$\text{i.e.,} \quad \epsilon = \frac{\omega}{c}(n'_y - n'_x)d, \tag{8.58}$$

$$= \frac{\omega}{c}n_o^3 r_{63} E_z d, \text{ (using Eqs. (8.43) and (8.44))}, \tag{8.59}$$

$$= \frac{\omega}{c}n_o^3 r_{63} V, \tag{8.60}$$

where $V = E_z d$ = voltage applied across the plate surfaces. We now calculate the polarization of the output wave. Assume that the input (incident) wave is linearly polarized and whose electric field vector in the laboratory (x, y) coordinate is horizontal, i.e., in Jones vector representation:

$$\boldsymbol{E}_i = \begin{bmatrix} 1 \\ 0 \end{bmatrix}. \tag{8.61}$$

We assume also that the modified (x', y') coordinates of the crystal make an angle of $45'$ with the (x, y) coordinates. Assuming that $\epsilon = \pi$, we can calculate the output polarization using Eqs. (7.93) and (8.61). This is equivalent to a half-wave plate (Section 7.9, Example 2), except that in this case, the relative phase shift ϵ is induced by the external field. From Eq. (7.106), the output polarization is rotated by $90°$ ($\because \psi = 45°$) (Fig. 8.2). The voltage at which $\epsilon = \pi$ is thus called the *half-wave voltage*, V_π, and by Eq. (8.60):

$$V_\pi = \frac{c\pi}{\omega n_o^3 r_{63}} = \frac{\lambda}{2n_o^3 r_{63}}. \tag{8.62}$$

In terms of V_π, Eq. (8.60) can be rewritten as

$$\epsilon = \pi V/V_\pi. \tag{8.63}$$

Thus, one can apply this technique to turn a laser beam's linear polarization by $90°$. Of course, if the angle is ψ, this electrically

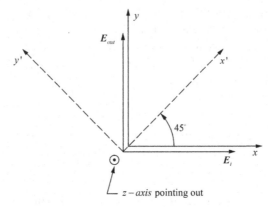

Fig. 8.2.   Polarization rotation by a half-wave voltage.

induced half-wave plate will turn the polarization by an angle of 2ψ (Sections 7.7 and 7.9).

Another voltage of interest is the *quarter-wave voltage* ($V_{1/4} \equiv V_{\pi/2}$) at which the crystal is equivalent to a quarter-wave plate (i.e., $\epsilon = \pi/2$). One sees that by Eq. (8.60), $V_{1/4} = 1/2 V_{\pi}$. Circular polarization will be obtained after a single pass through the EO crystal.

Case 2: AC field; amplitude modulation

We now vary the external field sinusoidally at the modulation frequency ω_m. Assume ω_m is much smaller than the optical frequency ω, i.e., $\omega_m \ll \omega$. We use a linear polarizer to analyze the output (Fig. 8.3). In general, the input light beam can be at any polarization state. We have chosen elliptical polarization in the figure. After passing through the AC field-biased z-cut KDP crystal, the polarization of the output is in a different general state, represented in the figure by another elliptical polarization; the polarizer transmits only the vertical component of the polarization. As the applied voltage changes with time, the output's polarization state changes so that the amplitude of its vertical component also changes with time. This leads to amplitude modulation.

In practice, one adds a quarter-wave plate between the output polarizer (or analyzer) and the KDP crystal (Fig. 8.4). The KDP is biased as in the discussion of Eq. (8.25) to (8.45). We note from Eqs. (8.43) and (8.44), for positive $E_{z'} n_{y'} > n_{x'}$, i.e., the velocity of the wave in the crystal whose polarization is in the y-direction

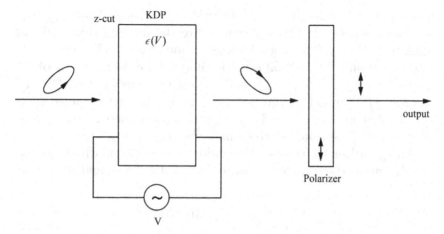

Fig. 8.3.   AC field modulation across a KDP EO crystal.

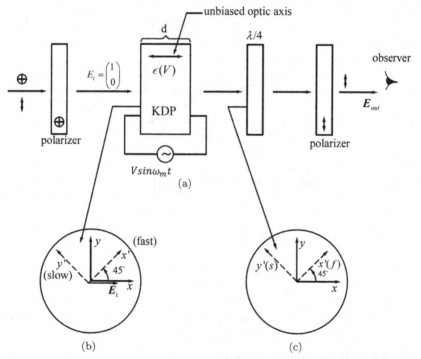

Fig. 8.4.   Electro-optic amplitude modulation of a light wave.

is smaller than that in the x'-direction ($\because v = \frac{c}{n}$; hence, greater n means smaller v). Thus, y' and x' are the induced slow and fast axes, respectively. The angle between x and x' is $45°$ (Eq. (8.30)). In Fig. 8.4(b) and (c), the following is chosen. The slow (s) axis of the quarter-wave plate is parallel to the slow (y') axis of the KDP and the fast (f) parallel to the fast (x'). Thus, the quarter-wave plate acts as if it is part of the KDP crystal with a fixed relative phase shift of $\pi/2$. The total relative phase shift is $\epsilon' = \epsilon + \pi/2$.

The polarizer transmits polarization in the vertical direction perpendicular to the z-direction as indicated. Its Jones matrix is from Eq. (7.96):

$$A \equiv A_y = e^{-i\phi} \begin{bmatrix} 0 & 0 \\ 0 & 1 \end{bmatrix}. \tag{8.64}$$

We can calculate $\boldsymbol{E}_{\text{out}}$ using Eqs. (7.93), (7.93′), (8.64) and (8.61):

$$W = AR(-45°)M_{\text{p}}R(45°), \tag{8.65}$$

where

$$R(45°) = \frac{1}{\sqrt{2}} \begin{bmatrix} 1 & 1 \\ -1 & 1 \end{bmatrix}, \tag{8.66}$$

$$R(-45°) = \frac{1}{\sqrt{2}} \begin{bmatrix} 1 & -1 \\ 1 & 1 \end{bmatrix}, \tag{8.67}$$

$$M_{\text{p}} = e^{-i\phi} \begin{bmatrix} e^{-i\epsilon'/2} & 0 \\ 0 & e^{i\epsilon'/2} \end{bmatrix}, \tag{8.68}$$

$$A = \begin{bmatrix} 0 & 0 \\ 0 & 1 \end{bmatrix}, \text{(omitting the phase)}, \tag{8.69}$$

$$\boldsymbol{E}_{\text{out}} = W\boldsymbol{E}_{\text{i}} = W \begin{pmatrix} 1 \\ 0 \end{pmatrix},$$

$$= i\sin(\epsilon'/2)e^{-i\phi} \begin{pmatrix} 0 \\ 1 \end{pmatrix}. \tag{8.70}$$

The power transmission is then

$$T = \frac{\boldsymbol{E}_{\text{out}}^* \cdot \boldsymbol{E}_{\text{out}}}{\boldsymbol{E}_{\text{i}}^* \cdot \boldsymbol{E}_{\text{i}}} = \frac{\sin^2(\epsilon'/2)}{1}$$

or

$$T = \sin^2\left(\frac{\epsilon'}{2}\right).$$ (8.71)

Now,

$$\epsilon' = \epsilon + \pi/2,$$ (8.72)

and if

$$\epsilon \equiv \epsilon_{\mathrm{m}}\sin\omega_{\mathrm{m}}t,$$ (8.73)

we obtain

$$T = \sin^2\left(\frac{\pi}{4} + \frac{\epsilon_{\mathrm{m}}}{2}\sin\omega_{\mathrm{m}}t\right),$$
$$= \left\{\frac{1}{2} - \cos\left[\frac{\pi}{2} + \epsilon_{\mathrm{m}}\sin\omega_{\mathrm{m}}t\right]\right\},$$
$$= \frac{1}{2}\left\{1 + \sin(\epsilon_{\mathrm{m}}\sin\omega_{\mathrm{m}}t)\right\}.$$ (8.74)

For $\epsilon \ll 1$,

$$T = \frac{1}{2}\left\{1 + \epsilon_{\mathrm{m}}\sin\omega_{\mathrm{m}}t - \cdots\right\},$$
$$= \frac{1}{2}\left(1 + \epsilon_{\mathrm{m}}\sin\omega_{\mathrm{m}}t\right).$$ (8.75)

The transmission follows ω_m, or the transmitted power is in phase with the modulation frequency. The amplitude ϵ_m is given by Eq. (8.60):

$$\epsilon_m = \frac{\omega}{c}n_o^3 r_{63}V,$$ (8.76)

where V is the amplitude of the applied voltage $V\sin\omega_m t$.

In terms of the half-wave voltage V_π (Eq. (8.62)):

$$\epsilon_{\mathrm{m}} = \frac{\pi V}{V_\pi}.$$ (8.77)

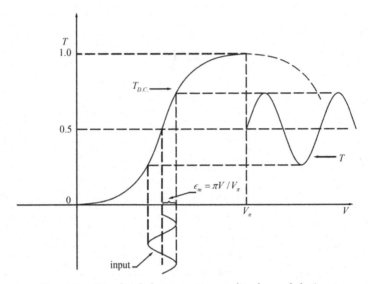

Fig. 8.5. Result of electro-optic amplitude modulation.

Equation (8.75) becomes:

$$T = \frac{1}{2}\left[1 + \frac{\pi V}{V_\pi}\sin\omega_m t\right]. \qquad (8.78)$$

Figure 8.5 shows such a modulated transmission with respect to the DC transmission:

$$T_{\text{D.C.}} = \sin^2(\epsilon/2), \quad (\text{using Eq. (8.71)}),$$

$$= \sin^2\frac{\pi}{2}\left[\frac{V}{V_\pi}\right], \quad (\text{using Eq. (8.63)}). \qquad (8.79)$$

Case 3: AC field: Phase modulation (FM)

The same z-cut KDP with the same bias voltage can be used, except that the input wave's polarization is parallel to the x'-axis of the crystal (Fig. 8.6). Since the input polarization is already parallel to one of the principal axes of the crystal (Fig. 8.6(b)), the state of polarization of the wave will remain the same in propagating through the crystal. However, its phase will be retarded. From Eq. (8.43), the expression for $n_{x'}$, which is the index of the wave whose polarization is in the x'-direction, we see that if $E_z = 0$, the incident wave sees the index n_o. Hence, the phase of the transmitted wave will change when E_z varies from zero to a finite value.

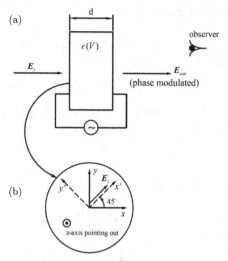

Fig. 8.6.   EO phase modulation.

If

$$\boldsymbol{E}_{\mathrm{i}} = \mathrm{A}\cos(\omega t - kz),\ \text{assuming no loss,}$$

$$\boldsymbol{E}_{\mathrm{out}} = \mathrm{A}\cos(\omega t - kz - k'd),$$

where

$$k'd = \frac{\omega}{\mathrm{c}} n_{x'} d,$$

$$= \frac{\omega}{\mathrm{c}} d \left(n_{\mathrm{o}} - \frac{1}{2} n_{\mathrm{o}}^3 r_{63} E_z \right) \quad \text{(by Eq. (8.43))},$$

$$= \frac{\omega n_{\mathrm{o}} d}{\mathrm{c}} - \frac{\omega n_{\mathrm{o}}^3 r_{63}}{2\mathrm{c}} V, \quad (V = E_z d).$$

If

$$V = V_{\mathrm{m}} \sin\omega_{\mathrm{m}} t,$$

$$\boldsymbol{E}_{\mathrm{out}} = \mathrm{A}\cos\left[\omega t - kz - \frac{\omega n_{\mathrm{o}} d}{\mathrm{c}} + \delta \sin\omega_{\mathrm{m}} t \right], \tag{8.80}$$

where

$$\delta \equiv \frac{\omega n_{\mathrm{o}}^3 r_{63}}{2\mathrm{c}} V_{\mathrm{m}}. \tag{8.81}$$

Equation (8.80) is an equation of phase modulation.

8.4   Quadratic EO Effect

We start with Eq. (8.10), the Taylor expansion of the change of impermeability due to an external field:

$$\Delta\eta_{ij}(\boldsymbol{E}) = r_{ijk}E_k + s_{ijkl}E_kE_\ell + \cdots \tag{8.82}$$

If, because of some reason, such as inversion symmetry in the crystal, $r_{ijk} = 0$, we then consider the contribution of the quadratic term in Eq. (8.82),
    i.e.,

$$\Delta\eta_{ij}(\boldsymbol{E}) = \eta_{ij}(\boldsymbol{E}) - \eta_{ij}(0),$$
$$= s_{ijkl}E_kE_\ell, \tag{8.83}$$

where from Eq. (8.9):

$$s_{ijkl} \equiv \frac{1}{2}\left[\frac{\partial^2\eta_{ij}}{\partial E_k\partial E_\ell}\right]_{E_k=E_\ell=0}. \tag{8.84}$$

Since the order of the partial differentiation is not important, we can interchange k and ℓ.
    Thus,

$$s_{ijk\ell} = s_{ij\ell k}. \tag{8.85}$$

Also, because of symmetry of the index tensor $\epsilon_{ij} = \epsilon_{ji}$, and thus $\eta_{ij} = \eta_{ji}$ (Eq. (8.14)).
    Equation (8.84) is unchanged with i and j interchanged,
    i.e.,

$$s_{ijk\ell} = s_{jik\ell}. \tag{8.86}$$

Because of Eqs. (8.85) and (8.86), we are allowed to use the following short hand notation.

(ij) or $(k\ell)$	Short Hand Notation
(11)	1
(22)	2
(33)	3
(23)=(32)	4
(13)=(31)	5
(12)=(21)	6

The index ellipsoid (Eq. (8.6))

$$\eta_{ij}(\boldsymbol{E})x_i x_j = 1$$

now becomes (using Eq. (8.83)):

$$\eta_{ij}(0)+s_{ijk\ell}E_k E_\ell x_i x_j = 1. \qquad (8.87)$$

From now on, (x, y, z) and (x_1, x_2, x_3) are freely interchanged whenever convenient. Equation (8.87) becomes:

$$\frac{x^2}{n_x^2} + \frac{y^2}{n_y^2} + \frac{z^2}{n_z^2} + s_{ijk\ell}E_k E_\ell x_i x_j = 1. \qquad (8.88)$$

Using the short hand notation and expanding the summation, the fourth term in Eq. (8.88), after some tedious but straight forward algebra (<u>exercise</u>), becomes

$$s_{ijk\ell}E_k E_\ell x_i x_j = (x_1^2, x_2^2, x_3^2, 2x_2 x_3, 2x_1 x_3, 2x_1 x_2)$$

$$\times \begin{bmatrix} s_{11} & s_{12} & s_{13} & s_{14} & s_{15} & s_{16} \\ s_{21} & s_{22} & s_{23} & s_{24} & s_{25} & s_{26} \\ s_{31} & s_{32} & s_{33} & s_{34} & s_{35} & s_{36} \\ s_{41} & s_{42} & s_{43} & s_{44} & s_{45} & s_{46} \\ s_{51} & s_{52} & s_{53} & s_{54} & s_{55} & s_{56} \\ s_{61} & s_{62} & s_{63} & s_{64} & s_{65} & s_{66} \end{bmatrix} \begin{bmatrix} E_1^2 \\ E_2^2 \\ E_3^2 \\ 2E_3 E_2 \\ 2E_1 E_3 \\ 2E_1 E_2 \end{bmatrix}$$

$$\equiv M. \qquad (8.89)$$

The ellipsoid for the second-order (quadratic) EO effect is (from Eq. (8.88)):

$$\frac{x^2}{n_x^2} + \frac{y^2}{n_y^2} + \frac{z^2}{n_z^2} + M = 1, \qquad (8.90)$$

where M is given by Eq. (8.89).

Example: *An isotropic medium: Kerr effect*

We consider an isotropic medium containing polarizable molecules such as liquid CS_2. The CS_2 polar molecules can be aligned in an electric field. Of course, not all the CS_2 molecules are aligned along the field direction. Only a tiny fraction of the molecules are either lined up or partially lined up, the rest being "destroyed" randomly by collisions. The result is the creation of a net field-induced polarization in the liquid in the direction of the applied field. This makes the liquid anisotropic. The quadratic EO coefficient or matrix s_{ij} (Eq. (8.89)) of an isotropic medium can be found from tabulated sources, e.g., (Yariv and Yeh, 1984). Thus, we have

$$M = (x_1^2, x_2^2, x_3^2, 2x_2x_3, 2x_1x_3, 2x_1x_2)$$

$$\times \begin{bmatrix} s_{11} & s_{11} & s_{12} & 0 & 0 & 0 \\ s_{12} & s_{11} & s_{12} & 0 & 0 & 0 \\ s_{12} & s_{12} & s_{11} & 0 & 0 & 0 \\ 0 & 0 & 0 & \frac{1}{2}(s_{11} - s_{12}) & 0 & 0 \\ 0 & 0 & 0 & 0 & \frac{1}{2}(s_{11} - s_{12}) & 0 \\ 0 & 0 & 0 & 0 & 0 & \frac{1}{2}(s_{11} - s_{12}) \end{bmatrix}$$

$$\times \begin{bmatrix} E_1^2 \\ E_2^2 \\ E_3^2 \\ 2E_3E_2 \\ 2E_1E_3 \\ 2E_2E_1 \end{bmatrix} \tag{8.91}$$

and the index ellipsoid is:

$$\frac{x^2}{n^2} + \frac{y^2}{n^2} + \frac{z^2}{n^2} + M = 1. \tag{8.92}$$

Note the isotropic index $n = $ constant.

Special case: $\boldsymbol{E} = E_3\hat{z}$,

i.e., $E_1 = E_2 = 0$.

On simplifying Eq. (8.91), Eq. (8.92) becomes

$$x_1^2\left\{\frac{1}{n^2} + s_{12}E_3^2\right\} + x_2^2\left\{\frac{1}{n^2} + s_{12}E_3^2\right\} + x_3^2\left\{\frac{1}{n^2} + s_{11}E_3^2\right\} = 1. \tag{8.93}$$

$$\text{Let} \quad n_o^2 = \left[\frac{1}{n^2} + s_{12}E_3^2\right]^{-1}, \tag{8.94}$$

$$n_e^2 = \left[\frac{1}{n^2} + s_{11}E_3^2\right]^{-1}. \tag{8.95}$$

Equation (8.93) becomes

$$\frac{x^2}{n_o^2} + \frac{y^2}{n_o^2} + \frac{z^2}{n_e^2} = 1.$$ (8.96)

This is an equation for a uniaxial medium. Equations (8.94) and (8.95) can be simplified to yield expressions of n_o and n_o under the assumption of small index change. The derivation is similar to that of Eqs. (8.41) and (8.42) and is left as an <u>exercise</u>. The result is

$$n_o = n - \frac{1}{2}n^3 s_{12} E_3^2,$$ (8.97)

$$n_e = n - \frac{1}{2}n^3 s_{11} E_3^2.$$ (8.98)

From the form of Eq. (8.96) and the knowledge gained in Chapter 6 when considering optical anisotropy, we conclude that the induced ordinary (o) and extraordinary (e) waves in the normally isotropic liquid (of CS_2) have orthogonal polarizations. Such liquid in a cell with plane windows acts as a wave plate. The relative phase shift of the two waves in traversing the liquid is given by Eq. (7.32), in which the index difference, or *birefringence* is (using Eqs. (8.97) and (8.98)):

$$n_e - n_o = \frac{1}{2}n^3(s_{12} - s_{11})E_3^2.$$ (8.99)

This is often expressed as

$$n_e - n_o = K\lambda E^2,$$ (8.100)

where

$$\lambda \equiv \text{vacuum wavelength of the light}$$
$$K \equiv \text{Kerr constant}$$

The values of K of some practical media can be found in tabulated sources, e.g., Yariv and Yeh (1984).

8.5 EO Shutter: Short Laser Pulse Slicer and 0-svd.tchincr Lasers

The basic idea of using a shutter is simply opening and closing the shutter at precise moments so that only during the open phase of the shutter will light pass through it. For high-speed events, one should be able to open and close the shutter in as short a time as possible so as to transmit only a very short light pulse. A mechanical shutter is too slow.

The reasons why one needs to have very short pulses, especially laser pulses, are manifold. From the point of view of generating very intense instantaneous power of laser radiation, it is easy to arrive at a very high peak power using short pulses because the power is defined as energy/pulse duration. For a constant energy content, the shorter the laser pulse is, the higher the peak power will be. From the point of view of data processing and communication, short pulses permit us to increase the data sending rate. From the point of view of observing phenomena that last for only a very short time, we need short pulses to illuminate and observe the phenomena, similar to using an extremely fast stroboscope. From the point of view of "instantaneous" excitation of matter, short pulses are required. Finally, just for the sake of pushing back the frontier of ultrashort pulse technology, one would like to produce the shortest pulses ever. The shortest laser pulse at around 0.62 μm is about a few femtoseconds (1 femtosecond = 10^{-15} seconds). A new question is now posed: "How does an extremely short (femtosecond) and extremely intense laser pulse interact with matter?" Could there be new physics discovered? Such are some research problems at present concerning intense short pulses.

In this section, we make use of the fact that we can change the state of polarization of a laser or light beam in an EO crystal and try to turn it into a fast shutter (<1 nanosecond or 10^{-9} sec at the limit). We again use the same KDP as an example (Section 8.3). Two cases will be discussed qualitatively. The KDP shutter is either inside or outside the laser cavity.

Case 1: Extra-cavity electro-optic shutter: Slicing short laser pulses

Figure 8.7 shows the schematic setup. The KDP crystal is placed between two crossed polarizers. The one at the left (incident side)

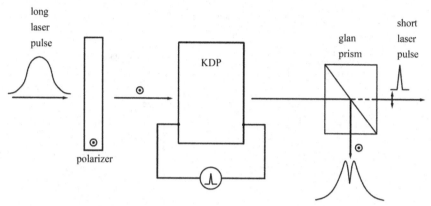

Fig. 8.7.   Electro-optic laser pulse slicer (principle).

is not necessary if the long laser pulse is already polarized in the direction allowed by the polarizer. The uniaxial KDP's optic axis is along the propagation axis, as in the case of Section 8.3. When a long laser pulse polarized perpendicular to the paper arrives at the crystal, no high voltage is applied to the KDP yet. Thus, the laser light passes without any change in the state of polarization because the laser propagates along the optic axis. This light, when reaching the Glan prism (or Glan Thompson prism or a stack of plates of isotropic material set at the Brewster angle (Chapter 3, etc.)), will be reflected entirely and the transmission is zero. When the pulse's maximum passes through the KDP, a high voltage pulse of peak voltage equal to V_π (Section 8.3), the half-wave voltage, and of very short duration (say, 1 ns) is applied across the crystal. During this brief period of time, the KDP acts as a half-wave plate, and the polarization of the transmitted laser pulse is turned 90° with respect to the incident polarization so that the Glan prism now transmits the pulse. After the short voltage pulse is over, the laser is again unperturbed and is completely reflected by the Glan prism. In Fig. 8.7, the form of the reflected and transmitted pulses is shown.

The circuit for the creation of the short high voltage pulse is not easy to understand, although the circuit design is simple. Figure 8.8 shows a circuit using the technique of a laser-triggered spark gap (LTSG). Figure 8.8(a) shows a schematic drawing of a real laboratory setup, while Fig. 8.8(b) is the abbreviated circuit diagram of (a). The reader should have knowledge of electric (voltage) wave propagation

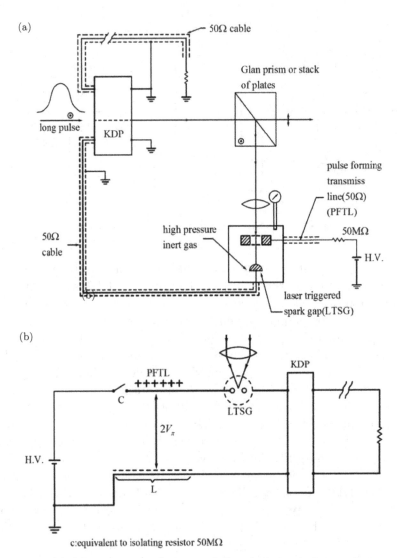

Fig. 8.8. (a) Schematic optic layout and electrical circuit for an electro-optic laser pulse slicing system. (b) Transmission line analysis of an EK laser pulse slicing system.

in transmission lines in order to appreciate fully what is described here (cf. any textbook on applied electromagnetism for engineers).

Referring to Fig. 8.8(b), the high voltage source (HV) is used to charge up the pulse forming 50H transmission line (PFTL) to a voltage of $2V_\pi$ (V_π is the half-wave voltage for the crystal KDP.) After this, the contact C is open, isolating the PFTL from the HV. A laser pulse arrives some time later and is focused into the LTSG, creating plasma inside the cell (spark gap). The rise time of the plasma formation is very fast (<1 ns). Now the PFTL "discharges" across the remaining part of the circuit. A high voltage "square" wave of peak voltage V_π and duration $2L/v$ is created, where L is the length of the PFTL and v is the wave propagation velocity in the cable. The pulse travels across the KDP and terminates at the 50Ω cables, of course. In principle, any cable can be used so long as the terminating resistance is equal to the impedance of the cable. The terminating cable after the KDP crystal is very long because one wants to avoid residual pulse reflection arriving at the KDP before the short pulse is over. When this pulse travels across the KDP crystal the latter behaves like a half-wave plate. The laser polarization passing through it during this time will be turned 90° and is transmitted by the Glan prism, see Fig. 8.8(a). Before and after, all laser light that passes through the KDP crystal along its optic axis will be rejected by the Glan prism.

The duration of the electrical pulse, being $2L/v$, can be varied by changing the length of the pulse forming cable (PFTL), although, in principle, one can shorten the cable indefinitely, at least two difficulties limit the pulse duration to go shorter than < 1 ns. The first is mechanical, i.e., it is not so easy to connect a very short cable in a practical system. But this is secondary. The more important difficulty is the rise time of the pulse, which is controlled by the rise time of the plasma formation in the spark gap. The latter is at best ~0.05 − 0.1 ns so that the voltage pulse cannot be shorter than about 0.1 ns. But in reality, stray capacitance, including that of the crystal, limits the rise time to ~1 ns. The pulse-forming cable for such a duration is still long enough (order of a few cm) for easy connection. The resulting laser pulse transmitted by the Glan prism is, of course, of the same duration as the voltage pulse and thus is limited to ~1 ns.

Case 2: Intracavity electro-optic shutter and Q-switching a laser

We show here two common intracavity techniques of switching out intense short (~10 ns) laser pulses. One uses a quarter-wave voltage across the EO crystal, the other, a half-wave voltage. Such techniques are called Q-switching.

Before going into them, let us define *Q-switching* and its principle.

Definition: *Q-switching*

Because Q or the Q of the cavity (see Chapter 4, Eq. (4.4)) represents the loss in the cavity, Q-switching means switching the cavity from a high loss state (low Q) that prevents laser oscillation to a low loss state that favors oscillation. The word *O-spoiling* was also used in older literature.

Definition: *Giant pulse*

The laser pulse coming out of a Q-switched laser, being very intense (a lot of energy in a very short time) is sometimes called a giant pulse. A few hundred millijoules in about 10 ns from a Q-switched Nd:YAG laser oscillator of beam diameter 5 mm is about the maximum before laser damage of optical components in the cavity takes place. (Hence, one needs amplification by first expanding the beam diameter using telescopes and spatial filters.)

Principle of Q-switching process

The basic idea is the following. Let us insert an optical element or a group of optical elements inside the cavity. The property of such an element (or group of elements) is to prevent the laser cavity from oscillating at low photon flux inside the cavity. That is, it represents a high loss (low Q). Note: In Chapter 2, the Q or quality factor is defined as the inverse of the loss (Eq. (2.4″)). Now, we start pumping the active medium inside the cavity. In Fig. 8.9, light propagating into B is generated from the active medium A through spontaneous emission plus stimulated emission in A after at most one round trip, as shown by the U-arrowed path. But B has a high loss by definition and rejects this light, either through absorption or scattering or reflection. Thus, there is no feedback, and the tendency of oscillation is blocked.

Meanwhile, the pump source keeps on pumping A, increasing its population inversion above the normal threshold lasing level. At the same time, light waves, one after another (or wave after wave) emanating from A hit B with stronger and stronger intensities because the population inversion keeps increasing, so that each successive wave will see a higher gain. Yet they are all absorbed, until the inversion becomes very much above the threshold. Then, "suddenly", B

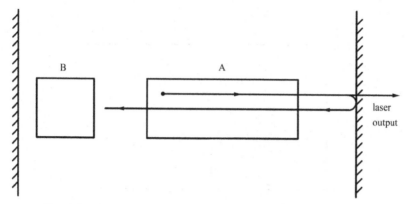

Fig. 8.9. Q-switching schematic with a saturable absorber B.

lets go an incoming strong wave of light because it cannot "stand" the intensity anymore, and the wave starts bouncing back and forth between the two mirrors. Because of the very high population inversion, the gain is very high, so that the radiation density builds up extremely fast. In the time for just a few round trips, the population inversion is completely depleted, and the output is a pulse containing a lot of energy in a very short time ($\sim$10 ns). This is a giant pulse. If by this time, the pumping is also terminated (or nearly so), there is only one pulse in the output.

(A) Passive O-switching

In *passive Q-switching*, the element B in Fig. 8.9 is a *saturable absorber* that could be a solid or liquid or gas depending on the wavelength of the laser and the availability of appropriate materials. The principle of operation is the same as before. B absorbs all weak radiation coming from A while the latter builds up its population inversion. When the inversion reaches a certain high level, the next wave of light going towards B from within A is sufficiently amplified (in at most two passes), reaching an intensity called the *saturation intensity*, I_S, of the absorber. This forces B to let go of the wave as if B is punched through by the strong wave. Figure (8.10) illustrates the transmission characteristics of a saturable absorber. Oscillation is started, and a giant pulse is created.

We now consider the EO techniques.

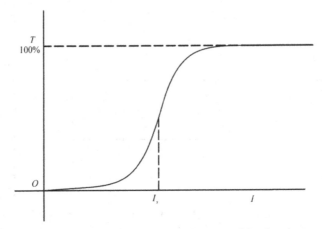

Fig. 8.10.   Transmission curve of a saturable absorber.

(B) Quarter-wave voltage Q-switching

Figure 8.11 shows the technique of using a quarter-wave voltage —
(a) shows the schematic setup. The same KDP crystal is set between
the totally reflecting end mirror of the laser cavity and a Glan prism.
The high voltage $V_{1/4}$ across the KDP is turned on at $t = 0$ when the
laser fires; (b) shows the path of a light beam coming from the active
medium after the voltage across the KDP is turned on. Assuming
the light beam is randomly polarized, let us call the linear polarization perpendicular to the page the P-polarization and the other
($\updownarrow$), V-(vertical) polarization. The P-polarization part of the light is
rejected by the Glan prism while the V-polarization is transmitted.
After passing once through the KDP towards the end mirror, the
V-polarization is transformed into circular polarization because the
KDP acts as a quarter-wave plate at the bias voltage of $V_{1/4}$.
The beam is then reflected, still circularly polarized, and passes again
through the KDP with the voltage $V_{1/4}$ still on. The circular polarization is now turned into a P-polarization. The reader should use either
the Jones matrix or the analytical method to show that such double
passage through a quarter-wave plate produces indeed a rotation of
the linear polarization by 90° (i.e., exercise). This P-polarization is
again reflected by the Glan prism. (The above means that the Glan
prism and the KDP together constitute the element B in Fig. 8.9.)

The above situation discourages any laser oscillation from occur-
ring. Meanwhile, the pumping of the active medium keeps on going,

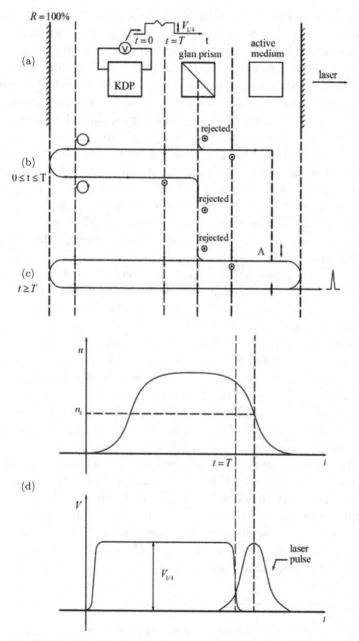

Fig. 8.11.   Quarter-wave voltage Q-switching of a laser.

and the population inversion builds up significantly above the threshold value n_t (see Fig. 8.11(d), top). The voltage pulse evolution $V(t)$ is shown in the bottom part of (d). Soon after the population inversion reaches the maximum, the HV pulse is terminated at $t = T$, and the laser starts oscillating as shown in (c). Now, the KDP is inactive so that the V-polarization that passes through it is unaffected since the propagation is along the KDP's optic axis. The Glan prism transmits the V-polarization, which then propagates through the active medium and is amplified. The reflected beam from the front mirror, after being amplified again by the active medium will pass through the Glan prism, thus closing the "circle" of oscillation. During this time, the pumping is significantly reduced or near zero.

Since the active medium is highly inverted, the amplification of light is very significant in every pass, thus depleting the inversion quickly. After a few round-trip passes, the inversion is depleted to below the threshold value. (No replenishment of the inversion is possible now because the pumping is nearly zero by design.) The laser pulse transmitted through the output-coupling mirror is at its maximum at the time when the rate of change of the inversion $\frac{dn}{dt}$ is still very large. It can be shown that this occurs at $n = n_t$, the threshold inversion. The bottom part in (d) shows the laser pulse. It shows also that even after n_t is reached, the strong radiation pulse that has been quickly released is still "sucking" energy out of the remaining inversion in the active medium. This is because the radiation in the cavity is now so strong that it takes very little time to "clean up" the remaining inversion through stimulated emission. The time is so short that the passive loss in the cavity, being a statistical quantity (i.e., average over a long time), did not fully affect the gain. (Note that the threshold inversion is in reality defined as a balance between absolute inversion and passive statistical loss. If the depletion of the inversion is very fast, the laser radiation does not see the full passive loss, i.e., the loss is effectively reduced. Hence, amplification continues beyond $n = n_t$ for a short while.)

(c) Half-wave voltage Q-switching

Figure 8.12 shows a similar analysis as in Fig. 8.11. It is left as an <u>exercise</u> for the reader to follow the figure and analyze it accordingly.

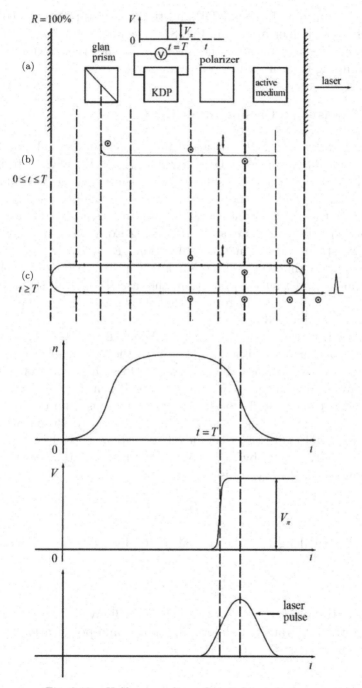

Fig. 8.12.   Half-wave voltage Q-switching of a laser.

We just point out that the KDP and the two cross polarizers constitute the Q-switch element B in Fig. 8.9. The high voltage V_π is turned on at time $t = T$ soon after the inversion reaches the maximum and the pumping is significantly reduced.

8.6 Transverse Biasing of an EO Crystal

So far, in Section 8.5, for pedagogical reasons, we have always used the longitudinally biased KDP (i.e., $\boldsymbol{E} = E_3 z^\wedge$, Eq. (8.25)) to illustrate modulation, the rapid shutter (pulse slicer) and Q-switching. In general, the crystal does not have to be longitudinally biased in order to achieve the above applications. It could be transversely biased, as mentioned briefly in the example following Eq. (8.46′), in which the KDP crystal is biased in the x-direction ($\boldsymbol{E} = E_1 \hat{x}$). (We always keep in mind that the z-direction is the propagation direction by definition.) Under such a condition, the laser beam will still see the electrically induced anisotropy so that its phase and/or polarization state will be changed.

To see this, we consider Eq. (8.46). When $\boldsymbol{E} = 0$, Eq. (8.46) gives the ellipsoid shown in Fig. 8.13(a). With the propagation direction **k** in the z-axis, the laser beam passing through it will not experience any change in phase or polarization. When $\boldsymbol{E} = E_1 \hat{x} \neq 0$, the ellipsoid becomes tilted (rotated about the x-axis) and deformed as given by Eq. (8.52) and illustrated in Fig. 8.13(b). Propagation in the z-axis will cause the laser beam to "see" a biaxial crystal, i.e., $n_{x'} = n_\mathrm{o}$ and n'_y given by Eq. (8.54). The phase difference of the two components in the crystal is

$$\varepsilon = \frac{\omega}{c}(n'_x - n'_y)\ell,$$

where ℓ is the length of the crystal. Using Eqs. (8.53) and (8.54)

$$\varepsilon = \frac{\omega\ell}{2c}n_\mathrm{o}^3 r_{41} E_3 \tan\alpha.$$

Thus controlling the length i and field E_3 will give us any phase difference that we want and, hence, any phase and polarization change of the laser beam.

(a)

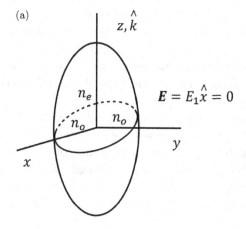

(b)

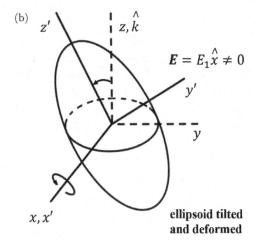

Fig. 8.13.   Transverse biasing of an EO crystal.

Many other EO crystals use transverse biasing, for example, CdTe and GaAs are both transversely biased for use with TEA-CO_2 laser beams, mostly for slicing pulses. We will not go into any details here.

8.7   Closing Remark

The EO effect is an electric field-induced anisotropy in materials. When we use this effect in different applications, the EO material functions essentially as an active wave plate — active in the sense that the electric field-induced anisotropy can be turned on and off at will, thus modifying the phase and polarization state of the transmitted light (laser) wave. When the EO material is used as a light modulator or fast shutter, one would certainly want to have it operate at very high frequencies or high speed, though there is a limit to the speed (or frequency response), of course.

From the basic design, when we apply a voltage across a pair of electrodes sticking on two surfaces of an EO medium, the ensemble becomes a capacitor. Thus, the capacitance of the EO cell, together with the impedance of the other electrical elements, constitutes a resultant impedance that limits the speed of the system. Since this becomes an electrical design problem, we will not go into further detail here.

Reference

A. Yariv and P. Yeh (1984). *Optical Waves in Crystals: Propagation and Control of Laser Radiation*. Wiley-Intersciences, New York.

Chapter 9

Mechanical Force Induced Anisotropy and Acousto-Optics

In nature, there is no ideally rigid body, i.e., all existing materials are deformable bodies. When we apply a mechanical force onto any medium in nature, it will, in general, induce a change in the body. This general change can be decomposed into three types of fundamental motions, namely, translation, rotation and strain. Translation and rotation represent the change of an ideally rigid body, whereas the strain represents the body's deformation with respect to its original state. Such a deformation (sometimes called "elastic change") can be put into use in modern optoelectronic applications. Qualitatively, imagine that a strain is created in an anisotropic medium. Its internal molecular and atomic structure is thus modified, even very slightly. Such a slight change is sufficient to induce change in the material density and thus index. It means that the original index ellipsoid of the medium is slightly modified, similar to the case of electric field-induced change of the index ellipsoid in Chapter 8. A laser (light) beam, being very sensitive, will feel the change in passing through the medium, thus changing its polarization state. One can take advantage of such a change in the polarization state of the laser (light) beam to generate modulation to the laser. Similar to the effect of an electric field, a strain induced in an isotropic medium will generate in it an anisotropy that can be used advantageously. There are many ways of creating strains in a medium. In modern optoelectronics applications, the most popular way is to create acoustic (elastic) waves; hence, acousto-optics (or opto-elastic effect).

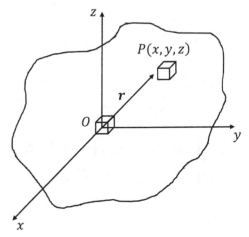

Fig. 9.1. Geometrical relationship between two elementary volumes at two neighboring points, O and P, inside a deformable body. The spatial Cartesian coordinate axis is fixed in space and the point O coincides with the origin before deformation.

9.1   The Strain Matrix

It is instructive to review briefly the derivation of the strain from the first principles. Consider, for ease of understanding, a solid body. (It could be liquid or gas.) We focus our attention on only two neighboring points, O and P, inside the body (Fig. 9.1). Let O initially coincide with the origin of a Cartesian coordinate system, which is fixed in space. The position of P is at $\boldsymbol{r} = (x, y, z)$. A mechanical force is applied to the body so that it has undergone a slight but general change. This means that there is a displacement of both P and O with respect to the origin. We denote the displacement of P by (ξ, η, ζ) and that of O by (ξ_O, η_O, ζ_O). The most general infinitesimal relative motion of P with respect to O is given by Taylor's expansion. (Note again that P is a neighboring point of O by definition so that the coordinates (x, y, z) of P are small quantities.)

$$\xi = \xi_O + \frac{\partial \xi}{\partial x}x + \frac{\partial \xi}{\partial y}y + \frac{\partial \xi}{\partial z}z + \text{higher-order terms}, \qquad (9.1)$$

$$\eta = \eta_O + \frac{\partial \eta}{\partial x}x + \frac{\partial \eta}{\partial y}y + \frac{\partial \eta}{\partial z}z + \text{higher-order terms}, \qquad (9.2)$$

$$\zeta = \zeta_O + \frac{\partial \zeta}{\partial x}x + \frac{\partial \zeta}{\partial y}y + \frac{\partial \zeta}{\partial z}z + \text{higher-order terms.} \qquad (9.3)$$

Because we have assumed only small changes, we can neglect the higher-order terms and rewrite Eqs. (9.1) to (9.3) in the following form:

$$\begin{pmatrix} \xi \\ \eta \\ \zeta \end{pmatrix} = \begin{pmatrix} \xi_O \\ \eta_O \\ \zeta_O \end{pmatrix} + \begin{pmatrix} 0 & \varphi_{12} & \varphi_{13} \\ \varphi_{21} & 0 & \varphi_{23} \\ \varphi_{31} & \varphi_{32} & 0 \end{pmatrix} \begin{pmatrix} x \\ y \\ z \end{pmatrix} + \begin{pmatrix} \varepsilon_{11} & \varepsilon_{12} & \varepsilon_{13} \\ \varepsilon_{21} & \varepsilon_{22} & \varepsilon_{23} \\ \varepsilon_{31} & \varepsilon_{32} & \varepsilon_{33} \end{pmatrix} \begin{pmatrix} x \\ y \\ z \end{pmatrix},$$

$$\qquad (9.4)$$

$$\equiv s_0 + s_1 + s_2, \qquad (9.5)$$

where

$$\varphi_{ik} \equiv \frac{a_{ik} - a_{ki}}{2} = -\varphi_{ki}, \quad i,k \equiv 1,2,3 \qquad (9.5')$$

$$\varepsilon_{ik} \equiv \frac{a_{ik} + a_{ki}}{2} = \varepsilon_{ki}, \quad i,k \equiv 1,2,3 \qquad (9.6)$$

(Thus, $\quad (a_{ik} = \varphi_{ik} + \varepsilon_{ik})) \qquad (9.7)$

$$a_{1k} \equiv \frac{\partial \xi}{\partial x_k} \ k= 1,2,3; \ (x_1, x_2, x_3) \equiv (x,y,z), \qquad (9.8)$$

$$a_{2k} \equiv \frac{\partial \eta}{\partial x_k} \ k= 1,2,3; \ (x_1, x_2, x_3) \equiv (x,y,z), \qquad (9.9)$$

$$a_{3k} \equiv \frac{\partial \zeta}{\partial x_k} \ k= 1,2,3; \ (x_1, x_2, x_3) \equiv (x,y,z). \qquad (9.10)$$

Note: (x, y, z) and (x_1, y_2, z_3) will be interchanged freely whenever convenient in what follows. (The reader should do the exercise of substituting the above definitions into Eq. (9.4) and obtaining Eqs. (9.1) to (9.3)). The first term on the right hand side of Eq. (9.4) represents a translation of P with respect to the origin fixed in space, i.e., s_0 in Eq. (9.5). The second term is the displacement s_1 of the tip of the vector r by a rotation through a small angle. This represents the rotation of the body. Let ϕ be the axial vector representing a rotation by an angle ϕ around an axis (Fig. 9.2), where

$$\phi \equiv \begin{pmatrix} \varphi_x \\ \varphi_y \\ \varphi_z \end{pmatrix} \qquad (9.11)$$

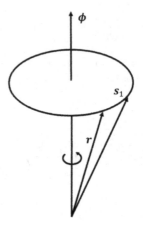

Fig. 9.2.   Relation between the axial vector and a rotation.

and where

$$\varphi_x \equiv \frac{a_{32} - a_{23}}{2}, \tag{9.12}$$

$$\varphi_y \equiv \frac{a_{13} - a_{31}}{2}, \tag{9.13}$$

$$\varphi_z \equiv \frac{a_{21} - a_{12}}{2}, \tag{9.14}$$

such that

$$\boldsymbol{s}_1 = \boldsymbol{\phi} \times \boldsymbol{r}. \tag{9.15}$$

As an <u>exercise</u>, the reader is asked again to substitute Eqs. (9.11) to (9.14) into (9.15) and obtain the second term on the right hand side of Eq. (9.4).

The third term on the right hand side of Eq. (9.4) is the displacement $\boldsymbol{s}_2$ caused by the strain (or deformation or elastic change.) The matrix (ε_{ik}) is the strain matrix (tensor). A slightly modified definition follows.

Consider the diagonal elements of the strain matrix.

$$\varepsilon_{11} = a_{11}, \quad \text{(by Eq. (9.6))},$$

$$= \frac{\partial \xi}{\partial x}, \quad \text{(by Eq. (9.8))}. \tag{9.16}$$

Similarly,

$$\varepsilon_{22} = \frac{\partial \eta}{\partial y}, \tag{9.17}$$

$$\varepsilon_{33} = \frac{\partial \zeta}{\partial z}. \tag{9.18}$$

They correspond to the increments in length per unit original length (or extensions) in the directions of x, y and z, respectively. The off diagonal terms can be similarly interpreted, but they represent the increment in one direction per unit original length in the other two directions. For example:

$$\varepsilon_{12} \equiv \frac{a_{12}+a_{21}}{2}, \quad \text{(from Eq. (9.6))},$$

$$= \frac{1}{2}\left(\frac{\partial \xi}{\partial y} + \frac{\partial \eta}{\partial x}\right), \quad \text{(from Eq. (9.8))},$$

$$2\varepsilon_{12} = \frac{\partial \xi}{\partial y} + \frac{\partial \eta}{\partial x}. \tag{9.19}$$

This type of combined "cross increments" (Eq. (9.19)) is defined as shearing and we define all "cross increments" of the type (9.19) as the shearing strains s_{ij}, or the shear. They were given by

$$s_{ij} \equiv 2\varepsilon_{ij}, \quad (i \neq j), \tag{9.20}$$

$$s_{ii} \equiv \varepsilon_{ii}, \quad (i = j). \tag{9.21}$$

Thus, a modified strain tensor (matrix) is given by:

$$s \equiv \begin{pmatrix} s_{11} & s_{12} & s_{13} \\ s_{21} & s_{22} & s_{23} \\ s_{31} & s_{32} & s_{33} \end{pmatrix}, \tag{9.22}$$

where $s_{ij}(i \neq j)$ is given by Eq. (9.20), and s_{kl} are given by Eqs. (9.16) to (9.18). The difference between (ε_{ij}) and Eq. (9.22) is simply Eq. (9.20). Since $\varepsilon_{ij} = \varepsilon_{ji}$ (from Eq. (9.6)):

$$s_{ij} = s_{ji}, \tag{9.23}$$

i.e., s is symmetric.

Example 1:

If there is no translation nor rotation, but only strain, and

$$s_2 = z \cdot A\cos(\Omega t - kz), \tag{9.24}$$

calculate the strain tensor.

Solution:

Since there is no translation or rotation, the first two terms on the right hand side of Eq. (9.4) are zero. We have

$$\begin{pmatrix} \xi \\ \eta \\ \zeta \end{pmatrix} = s_2. \tag{9.25}$$

Comparing Eqs. (9.24) and (9.25), we see that

$$\zeta = A\cos(\Omega t - kz), \tag{9.26}$$

$$\xi = \eta = 0,$$

i.e., ζ is a function of z. Thus, all the matrix elements of S are zero except the one containing $(\partial \zeta / \partial z)$. And this element is:

$$s_{33} \equiv \frac{\partial \zeta}{\partial z}, \text{ (from Eqs. (9.21) and (9.18))},$$

$$= AK \sin(\Omega t - kz) \tag{9.27}$$

and

$$s = \begin{pmatrix} 0 & 0 & 0 \\ 0 & 0 & 0 \\ 0 & 0 & AK \sin(\Omega t - kz) \end{pmatrix}. \tag{9.28}$$

Example 2:

Again, assuming no translation and rotation of a body, but the strain displacement s_2 is

$$s_2 = \hat{y} A \cos(\Omega t - Kz). \tag{9.29}$$

Find the strain matrix.

Solution:

Again, Eq. (9.25) is valid.

We have:

$$\xi = \zeta = 0, \tag{9.30}$$

$$\eta = A\cos(\Omega t - Kz), \tag{9.31}$$

i.e., η is a function of z and only elements of s containing $(\partial\eta/\partial z)$ is non-zero. This term is

$$s_{23} \equiv \frac{\partial\eta}{\partial z} + \frac{\partial\zeta}{\partial y},$$

$$= AK\sin(\Omega t - kz) + 0, \tag{9.32}$$

$$s = \begin{pmatrix} 0 & 0 & 0 \\ 0 & 0 & AK\sin(\Omega t - kz) \\ 0 & AK\sin(\Omega t - kz) & 0 \end{pmatrix}. \tag{9.33}$$

Note: When an acoustic wave is applied to a medium, it does not, in general, induce any rigid body displacement and rotation. Thus $s_0 = s_1 = 0$, unless the wave is extremely strong.

Before ending this section, we remind ourselves that O and P in Fig. 9.1 are two *neighboring* points. Thus, s_2 is a *local* resultant displacement of an infinitesimal volume due to the strain. An example is the propagation of an acoustic wave in a medium. Depending on the material structure, acoustic wave propagation induces at some local point a resultant *displacement* of a very small volume of the matter not necessarily *in the direction* of propagation of the acoustic wave. This displacement is s_2. If we define the propagation direction as the z-direction and the local "particle" (very small volume) displacement is $U(z,t)$, then

$$s_2 = U(z,t) = U(z,t)\hat{z}.$$

If $U(z,t) = A\cos(\Omega t - Kz)$, which is a propagating wave, we have:

$$s_2 = \hat{z}A\cos(\Omega t - Kz). \tag{9.34}$$

9.2 Mechanically Induced Anisotropy

We assume that some mechanical force has induced only a strain in a medium. Such a strain leads to a change in local densities and thus a change in the refractive index of the medium. Such index change is described by the change in the optical impermeability η_{ij} defined by Eq. (8.4) in Chap. VIII, i.e., $\eta_{ij} \equiv \varepsilon_0/\varepsilon_{ij}$. The change in η_{ij} can be expressed in terms of the strain tensor S because it is the strain that has induced the change in η_{ij}. Thus,

$$\triangle\eta_{ij} = p_{ijkl}s_{kl} + \text{higher-order terms} \,(i,\ j,\ k,\ l = 1, 2, 3), \qquad (9.35)$$

where repeated indices mean summation. s_{kl} are the matrix elements of S, and p_{ijkl} are the matrix elements of a proportionality matrix. In Eq. (9.35), it is assumed that $\triangle\eta_{ij}$ is proportional to the strain to the first order. Because there were higher-order terms that were neglected while defining strain in Section 9.1, we write down "higher-order terms" in Eq. (9.35). We will neglect them immediately because the first order term is generally small.

The physical meaning of Eq. (9.35) is that the change in each of the impermeability matrix (tensor) elements (i.e, $\triangle\eta_{ij}$, for the *(ij)*th element) is the combined consequences of each of the mechanically induced shear strain element S_{kl}; hence, the summation over k and l. Let $\eta_{ij}(s)$ be the *(ij)*th element of the perturbed impermeability and $\eta_{ij}(0)$ be the unperturbed one. Thus,

$$\triangle\eta_{ij} = \eta_{ij}(s) - \eta_{ij}(0). \qquad (9.36)$$

The perturbed index ellipsoid (cf. Eq. (8.6)) is:

$$\eta_{ij}(s)x_i x_j = 1. \qquad (9.37)$$

(Note again, repeated indices means summation.)
Using Eq. (9.36), this becomes:

$$\eta_{ij}(0)x_i x_j + \triangle\eta_{ij}x_i x_j = 1. \qquad (9.38)$$

The first term is again the unperturbed expression given by the left hand side of Eq. (8.5). The second term is:

$$\triangle\eta_{ij}x_i x_j = p_{ijkl}s_{kl}x_i x_j, \qquad (9.39)$$

$$\equiv M. \qquad (9.40)$$

We again use short hand notations. Because η_{ij} and s_{ij} are both symmetric tensors,

$$\text{i.e.,}\quad \eta_{ij} = \eta_{ji}, \quad \text{(from Eq. (8.14))}$$

and

$$s_{ij} = s_{ji}, \quad \text{(from Eq. (9.23))},$$

we can make the following simplification.

(ij) or (kl)	Short Hand Notation
(11)	1
(22)	2
(33)	3
(23) or (32)	4
(13) or (31)	5
(12) or (21)	6

From Eq. (9.39):

$$\triangle\eta_{ij} = \triangle\eta_m = p_{mn}s_n, \quad (m,n = 1,2,\cdots,6). \tag{9.41}$$

Note that in Eq. (9.41), (m,n) run from 1 to 6. Rewriting Eq. (9.38) using Eqs. (8.5) and (9.40):

$$\frac{x_1^2}{n_1^2} + \frac{x_2^2}{n_2^2} + \frac{x_3^2}{n_3^2} + M = 1, \tag{9.42}$$

where M can be calculated from expanding Eq. (9.41). This is left as an <u>exercise</u> to the reader. (This is similar to the case of Kerr EO effect, i.e., $p_{ijkl} \leftrightarrow s_{ijkl}$ and $s_{ij} \leftrightarrow E_k E_l$.)

$$M = (x_1^2, x_2^2, x_3^2, 2x_2 x_3, 2x_1 x_3, 2x_1 x_2)$$

$$\cdot \begin{pmatrix} p_{11} & p_{12} & p_{13} & p_{14} & p_{15} & p_{16} \\ p_{21} & p_{22} & p_{23} & p_{24} & p_{25} & p_{26} \\ p_{31} & p_{32} & p_{33} & p_{34} & p_{35} & p_{36} \\ p_{41} & p_{42} & p_{43} & p_{44} & p_{45} & p_{46} \\ p_{51} & p_{52} & p_{53} & p_{54} & p_{55} & p_{56} \\ p_{61} & p_{62} & p_{63} & p_{64} & p_{65} & p_{66} \end{pmatrix} \begin{pmatrix} s_1 \\ s_2 \\ s_3 \\ s_4 \\ s_5 \\ s_6 \end{pmatrix}. \tag{9.43}$$

The tensor elements of (p_{mn}) are known as *elasto-optic coefficients*, and the matrices for many useful materials are tabulated according to their symmetry. (See, for example, (Yariv and Yen, 1984)) We should recall that the index ellipsoid (as well as the Fresnel wave normals or ray surfaces) describes the *local response* of a medium, and every point in the medium has the same response (see Chapter 6). So does the strain (see the end of the previous section).

Example 1: *Acoustic wave propagating in water*

An acoustic wave propagates in water in the z-direction. The local particle (very small volume) displacement induced by the acoustic wave is

$$\boldsymbol{U}(z,t) = \hat{z}A\cos(\Omega t - Kz). \tag{9.44}$$

Find the index change.

Solution: Since s_2 is the local displacement (cf. last paragraph of Section 9.1):

$$s_2 = \boldsymbol{U}(z,t) = \hat{z}A\cos(\Omega t - Kz). \tag{9.45}$$

This is indentical to Eq. (9.24). Thus, from Eq. (9.28), the only non-zero matrix element of s is

$$s_{33} \equiv s_3, \quad \text{(by the short hand notation)},$$

$$= AK\sin(\Omega t - Kz). \tag{9.46}$$

Water is isotropic. The elasto-optic tensor (p_{mn}) is given by

$$[p_{mn}] = \begin{pmatrix} p_{11} & p_{12} & p_{12} & 0 & 0 & 0 \\ p_{12} & p_{11} & p_{12} & 0 & 0 & 0 \\ p_{12} & p_{12} & p_{11} & 0 & 0 & 0 \\ 0 & 0 & 0 & \frac{1}{2}(p_{11}-p_{12}) & 0 & 0 \\ 0 & 0 & 0 & 0 & \frac{1}{2}(p_{11}-p_{12}) & 0 \\ 0 & 0 & 0 & 0 & 0 & \frac{1}{2}(p_{11}-p_{12}) \end{pmatrix}, \tag{9.47}$$

$$s = \begin{pmatrix} 0 \\ 0 \\ s_3 \\ 0 \\ 0 \\ 0 \end{pmatrix}. \tag{9.48}$$

From Eq. (9.43), using Eqs. (9.47) and (9.48):

$$M = \left(x_1^2, x_2^2, x_3^2, 2x_2x_3, 2x_1x_3, 2x_1x_2 \right) \begin{pmatrix} p_{12}s_3 \\ p_{12}s_3 \\ p_{11}s_3 \\ 0 \\ 0 \\ 0 \end{pmatrix},$$

$$= p_{12}s_3x_1^2 + p_{12}s_3x_2^2 + p_{11}s_3x_3^2. \tag{9.49}$$

Substituting Eq. (9.49) into Eq. (9.42), using:

$$n_1 = n_2 = n_3 = n_0, \quad \text{(water is isotropic)},$$

$$x_1^2 \left(\frac{1}{n_o^2} + p_{12}s_3 \right) + x_2^2 \left(\frac{1}{n_o^2} + p_{12}s_3 \right)$$

$$+ x_3^2 \left(\frac{1}{n_o^2} + p_{11}s_3 \right) = 1. \tag{9.50}$$

Equation (9.50) is a uniaxial index ellipsoid if $p_{11} \neq p_{12}$, i.e., under the propagation of an acoustic wave, an isotropic medium becomes a uniaxial material in general. From Eq. (9.50)

$$(n_x^2)^{-1} = (n_y^2)^{-1} = \frac{1}{n_o^2} + p_{12}s_3 = \frac{1}{n_o^2} + p_{12}AK\sin(\Omega t - kz), \tag{9.51}$$

$$(n_z^2)^{-1} = \frac{1}{n_o^2} + p_{11}s_3 = \frac{1}{n_o^2} + p_{11}AK\sin(\Omega t - kz). \tag{9.52}$$

Under the assumption of small index change, an approximation can be made to Eqs. (9.51) and (9.52) in a similar way as that used in Eqs. (8.41) and (8.42). This is left as an <u>exercise</u>. The result is

$$n_x = n_y = n_o - \frac{1}{2}n_o^3 p_{12}AK\sin(\Omega t - kz), \tag{9.53}$$

$$n_z = n_o - \frac{1}{2}n_o^3 p_{11}AK\sin(\Omega t - kz). \tag{9.54}$$

Figure 9.3 shows the resultant index ellipsoid. From this figure, we see that if we send an electromagnetic (EM) wave of arbitrary polarization along the z-direction, the water will still behave isotropically because $n_x = n_y$. But n_x and n_y are both sinusoidal functions.

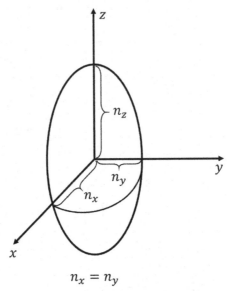

$$n_x = n_y$$

Fig. 9.3. An isotropic material becomes uniaxial under the propagation of an acoustic wave in the material. The figure shows the resultant index ellipsoid.

Freezing everything at a certain time t, both the polarization components $\boldsymbol{E}_x$ and $\boldsymbol{E}_y$ of the EM wave will see a medium of periodically varying index in the z-direction, i.e., each polarization (wave) sees an index grating. If the EM wave is polarized in the x-direction (i.e., $\boldsymbol{E}_y = 0$), it will see only the index grating (variation) of Eq. (9.53). Of course, if the EM wave propagates in an arbitrary direction, we should treat the whole affair as if it propagates in an anisotropic medium. There will be ordinary (o) and extraordinary (e) waves created. Each wave will see an index grating of the same Ω and K but of different amplitudes, because of Eqs. (9.53) and (9.54).

Example 2: Acoustic wave propagating in germanium

Germanium is also isotropic. However, its crystal symmetry dictates that if we propagate an acoustic wave in the direction of one of its crystal axes ($< 001 >$ direction, defined as z-axis), the shear strain induces a local particle displacement $\boldsymbol{s}_2$ in the y-direction (the $< 010 >$ direction of the crystal).

$$\boldsymbol{s}_2 = \widehat{y}A\cos(\Omega t - Kz). \tag{9.55}$$

This is identical to Eq. (9.29). From Eq. (9.33), the only non-vanishing element of the strain matrix is

$$s_4 \equiv s_{32} = s_{23}, \quad \text{(using short hand notation)},$$

$$= AK \sin(\Omega t - Kz). \tag{9.56}$$

The photo-elastic coefficients are, for Germanium

$$[p_{mn}] = \begin{pmatrix} p_{11} & p_{12} & p_{12} & 0 & 0 & 0 \\ p_{12} & p_{11} & p_{12} & 0 & 0 & 0 \\ p_{12} & p_{12} & p_{11} & 0 & 0 & 0 \\ 0 & 0 & 0 & p_{44} & 0 & 0 \\ 0 & 0 & 0 & 0 & p_{44} & 0 \\ 0 & 0 & 0 & 0 & 0 & p_{44} \end{pmatrix}, \tag{9.57}$$

$$s = \begin{pmatrix} 0 \\ 0 \\ 0 \\ s_4 \\ 0 \\ 0 \end{pmatrix}. \tag{9.58}$$

Equation (9.43) becomes:

$$M = (x_1^2, x_2^2, x_3^2, 2x_2x_3, 2x_1x_3, 2x_1x_2) \begin{pmatrix} 0 \\ 0 \\ 0 \\ p_{44}s_4 \\ 0 \\ 0 \end{pmatrix},$$

$$= 2p_{44}s_4x_2x_3 \tag{9.59}$$

and Eq. (9.42) becomes ($\because n_1 = n_2 = n_3 = n_o$, Ge being isotropic)

$$\frac{x_1^2}{n_o^2} + \frac{x_2^2}{n_o^2} + \frac{x_3^2}{n_o^2} + 2p_{44}s_4x_2x_3 = 1. \tag{9.60}$$

This equation is similar to Eq. (8.26). We need to make a rotation of the (x_1x_2) axes around the x_1-axis, similar to what was done to

Eq. (8.26). The result is that we obtain an ellipsoid along the new principal (x_1', x_2', x_3') axes after a rotation of $45°$ about the x_1-axis:

$$\frac{x_{1'}^2}{n_{x_{1'}}^2} + \frac{x_{2'}^2}{n_{x_{2'}}^2} + \frac{x_{3'}^2}{n_{x_{3'}}^2} = 1. \tag{9.61}$$

After an approximation similar to the first example (<u>exercise</u> again):

$$n_{x_1'} = n_0, \tag{9.62}$$

$$n_{x_2'} = n_o - \frac{1}{2} n_o^3 p_{44} AK \sin(\Omega t - kz), \tag{9.63}$$

$$n_{x_3'} = n_o + \frac{1}{2} n_o^3 p_{44} AK \sin(\Omega t - Kz). \tag{9.64}$$

The medium becomes biaxial because $n_{x_1'} \neq n_{x_2'} \neq n_{x_3'}$. At the same time, because the induced indices $(n_{x_2'}, n_{x_3'})$ vary periodically, an EM wave propagating through the medium will see a "grating" or periodic index variation. The polarization of the EM wave and the propagation direction determine the output characteristics.

Before ending this section, it should be stressed that any mechanical force acting on any matter will, in principle, generate some anisotropy in the matter so long as there is a strain in the material apart from rigid body translation and rotation. For instance, if one compresses a piece of glass (isotropic), it becomes uniaxial. If such kind of strained material is placed between two crossed polarizers, and a white light beam is sent through them, beautiful patterns representing the strain topography can be seen at the output. This is because the entrance polarizer transmits only light linearly polarized in its preferred direction. The polarization state of this transmitted light, after passing through the strained material, will be altered locally so that the exit polarizer (analyzer) transmits partially only the light that has passed through a strained region of the material.

9.3 Fundamentals of Acousto-optic Interaction

(a) Fundamentals of grating's diffraction

As mentioned in the previous section, when there is an acoustic wave propagating in a medium, an EM wave passing through the medium will see a 3D index grating. The EM wave will thus be diffracted by the grating. Although it is presumed that the readers are familiar with the formula of a diffraction grating, it is always instructive to quickly review the fundamentals of the subject.

The fundamental analysis of diffraction from a grating is the superposition of two EM waves scattered by two adjacent lines of the grating. In other words, we need only to know the consequence of the superposition of two scattered or diffracted waves (plane waves for simplicity) with the same propagation vector k and frequency ω, having undergone different optical paths. Their polarizations, assumed linear, are parallel. The two waves are:

$$E_1 = E_{o1} \cos(\omega t - k \cdot r_1), \tag{9.65}$$

$$E_2 = E_{o2} \cos(\omega t - k \cdot r_2). \tag{9.66}$$

The total field is:

$$E = E_1 + E_2 \tag{9.67}$$

and the intensity is

$$I \propto \langle E \cdot E \rangle. \tag{9.68}$$

We neglect the constant of proportionality:

$$I = \langle E \cdot E \rangle = \frac{1}{T} \int_0^T E \cdot E \, dt, \tag{9.69}$$

where

$$T \gg 2\pi/\omega. \tag{9.70}$$

Now,

$$\begin{aligned} E \cdot E &= (E_1 + E_2) \cdot (E_1 + E_2), \\ &= E_1^2 + E_2^2 + 2E_1 E_2 (\because E_1 \parallel a E_2). \end{aligned} \tag{9.71}$$

The cross term is the interference term and will be zero if E_1 is perpendicular to E_2, i.e., waves of perpendicular polarizations will not interfere.

From Eqs. (9.69) and (9.71):

$$I = E_1^2 + E_2^2 + 2E_1E_2,$$

$$= I_1 + I_2 + 2 < E_1E_2 >, \tag{9.72}$$

$$E_1E_2 = E_{01}E_{02}\cos(\omega t - \boldsymbol{k} \cdot \boldsymbol{r}_1)\cos(\omega t - \boldsymbol{k} \cdot \boldsymbol{r}_2)$$

$$= \cdots (\underline{\text{exercise}}),$$

$$= E_{01}E_{02}(\cos^2\omega t \cos \boldsymbol{k} \cdot \boldsymbol{r}_1 \cos \boldsymbol{k} \cdot \boldsymbol{r}_2$$

$$+ \sin^2\omega t \sin \boldsymbol{k} \cdot \boldsymbol{r}_1 \sin \boldsymbol{k} \cdot \boldsymbol{r}_2$$

$$- \sin \omega t \cos \omega t \sin \boldsymbol{k} \cdot \boldsymbol{r}_2 \cos \boldsymbol{k} \cdot \boldsymbol{r}_1$$

$$- \sin \omega t \cos \omega t \sin \boldsymbol{k} \cdot \boldsymbol{r}_1 \cos \boldsymbol{k} \cdot \boldsymbol{r}_2),$$

$$\left\langle \begin{bmatrix} \cos^2\omega t \\ \sin^2\omega t \end{bmatrix} \right\rangle = \cdots (\underline{\text{exercise}}),$$

$$= \frac{1}{2},$$

$$\langle \sin \omega t \cos \omega t \rangle = \cdots .(\underline{\text{exercise}}),$$

$$= 0,$$

$$\therefore \langle E_1E_2 \rangle = (\underline{\text{exercise}}),$$

$$= \frac{E_{O1}E_{O2}}{2}\cos[\boldsymbol{k} \cdot (\boldsymbol{r}_1 - \boldsymbol{r}_2)]. \tag{9.73}$$

Let

$$\varepsilon \equiv \boldsymbol{k} \cdot (\boldsymbol{r}_1 - \boldsymbol{r}_2), \tag{9.74}$$

$$\equiv phase\ difference,$$

$$\therefore < E_1E_2 > \equiv \frac{E_{O1}E_{O2}}{2}\cos\varepsilon, \tag{9.75}$$

$$\therefore I = I_1 + I_2 + E_{O1}E_{O2}\cos\varepsilon,$$

$$(\text{from Eqs. (9.72) and (9.75)}),$$

$$= \begin{cases} \text{maximum when } \cos\varepsilon \\ \quad = +1(\text{constructive interference}) \\ \text{minimum when } \cos\varepsilon \\ \quad = -1(\text{destructive interference}) \end{cases},$$

$$= \begin{Bmatrix} \text{maximum when } \varepsilon = 2m\pi \\ \text{minimum when } \varepsilon = (2m+1)\pi \end{Bmatrix}, \tag{9.76}$$

where

$$m = 0, 1, 2, \ldots$$

If $r_1 \parallel r_2$ and $\mathbf{k}$ are all in the z-direction, we have:

$$\varepsilon = k(z_1 - z_2),$$

$$= k\Delta l, \quad (\Delta l \equiv \text{optical path difference}),$$

$$= \frac{2\pi}{\lambda}\Delta l.$$

Hence,

$$I = \begin{cases} \text{maximum when } \frac{2\pi}{\lambda}\Delta l = 2m\pi \\ \text{minimum when } \frac{2\pi}{\lambda}\Delta l = (2m+1)\pi \end{cases},$$

$$= \begin{cases} \text{maximum when} \Delta l = m\lambda \\ \text{minimum when} \Delta l = \frac{2m+1}{2}\lambda \end{cases}. \tag{9.77}$$

This result will be used in the following subsection.

(b) Diffraction by a stationary acoustic grating in an isotropic medium

Assume that an acoustic wave is launched in an isotropic medium (e.g., water) and that the local particle displacement is:

$$U_Z \sim \hat{z}\cos(\Omega t - Kz).$$

(See Section 9.2, Example 1).

A sinusoidal index grating is thus generated in the medium. Let the incident plane EM wave be linearly polarized in the x-direction and the wave vector $\mathbf{k}$ in the $y - z$ plane as shown in Fig. 9.4(a). The index ellipsoid represents that of a uniaxial material in which $\mathbf{D}_0$ is always perpendicular to the $y - z$ plane (cf. Fig. 6.12). Hence, $\mathbf{D}_0 \parallel \mathbf{E}_x$. The polarization $\mathbf{E}_x$ thus sees sinusoidal index grating planes propagating in the z-direction (see Section 9.2, Example 1) since $\mathbf{D}_0$ sees the same thing. Since acoustic wave propagation velocity is much smaller than the speed of light ($v_{acoustic}/c \sim 10^{-5}$), we can consider the acoustic index grating as stationary in a first approximation. Such a grating is shown as horizontal lines in Fig. 9.4(b). We assume that the scattered light wave is also plane, which implies that the acoustic grating surfaces are plane and the interaction region is

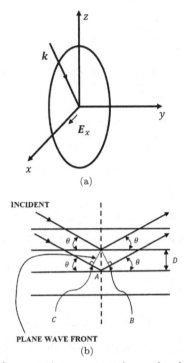

Fig. 9.4. Diffraction by a stationary acoustic grating in an isotropic medium. (a) Relationship between k, E_x of an incident EM wave and the resultant index ellipsoid of the material. (b) Acoustic diffraction.

large. As is well known, when a light beam crosses an interface separating two materials of indices n_1 and $n_2 (n_1 \neq n_2)$, Fresnel reflection takes place together with refraction (Chapter 3). The phenomenon is due to the difference in indices across the interface. Thus, in the case of acoustic plane index grating, there are periodic changes in the index. Higher index regions normally correspond to higher density regions of the same material. Light waves crossing these higher index regions will be reflected also. We idealize the periodic higher index regions as very thin planes (Fig. 9.4(b)). A plane wave incident onto two adjacent index planes is shown as two rays reflected by these index planes. (Refraction is ignored in our discussion.) The two reflected beams will be superimposed, giving rise to interference. Constructive interference takes place if the optical path difference between the two beams satisfies the first of Eq. (9.77),

$$\text{i.e.,} \quad \triangle l = m\lambda, (m = 1, 2, \dots) \tag{9.78}$$

or

$$AC + AB = m\lambda,$$

$$\because AC + AB = 2D \sin\theta,$$

$$\therefore 2D \sin\theta = m\lambda \tag{9.79}$$

or

$$2\pi \cdot \frac{2\sin\theta}{\lambda} = \left(\frac{m}{D}\right) \cdot 2\pi,$$

$$\text{i.e.,} \quad 2k \sin\theta = mK, \tag{9.80}$$

where $k = \frac{2\pi}{\lambda}$ is the EM wave vector and $K = \frac{2\pi}{D}$ is the acoustic wave vector. Two interpretations can be considered.

(i) We interpret Eq. (9.80) as the condition under which there is constructive interference between the two light beams reflected by two adjacent plane acoustic index grating planes whose wave vector is

$$mK = \frac{2\pi}{\left(\frac{D}{m}\right)} (m = 1, 2, 3, \ldots).$$

If $m \geq 2$, it means that higher harmonics (2nd, 3rd, etc.) of the acoustic wave of wavelength D/m have contributed to the constructive interference.

In the case of a pure sinusoidal acoustic wave, such harmonics are absent and $m = 1$.

Hence,

$$2k \sin\theta = K, \tag{9.81}$$

and we note that the wavelength λ in Eq. (9.79) is the wavelength of light in the medium, not in vacuum. $\lambda = \frac{v}{\upsilon} = \frac{\frac{c}{n}}{\upsilon} = \frac{1}{n}\lambda_o$,

$$\text{i.e.,} \quad \lambda_o = n\lambda, \tag{9.82}$$

where

$$n = \text{index of the material}$$

$$\lambda_o = \text{wavelength of light in vacuum.}$$

(ii) Another more classical interpretation of Eq. (9.79) or (9.80) is that m is the order of diffraction even if λ and D are the monochromatic wavelengths of the light and acoustic wave, respectively. But there is a constraint. Equation (9.79) can be re-written as

$$\sin\theta = \frac{m}{2}\frac{\lambda}{D}.$$

Therefore, either D cannot be too small or m cannot be too large. Otherwise, $\sin\theta$ will become greater than one, which is impossible. For high-frequency acoustic waves, D is small and is not much larger than λ. Thus, $m = 1$, or at best 2. In this section, we assume $m = 1$.

In summary, constructive interference occurs between light waves reflected by two adjacent acoustic index planes if Eqs. (9.79), (9.80) or (9.81) is satisfied. Since the distance D is the same between any pair of adjacent planes, all pairs of light waves reflected by different pairs of adjacent acoustic index planes will interfere constructively, and the combination of all these waves forms the diffracted wave with constructive interference. If D is kept constant and λ is changed, then θ has to be also changed in order to have constructive interference (i.e., satisfy Eq. (9.79), (9.80) or (9.81)).

Or if $\boldsymbol{k}$ (and thus θ and λ) is kept constant, one has to vary D (the acoustic wavelength) in order to obtain constructive interference in the reflection. Under the situation of constructive interference of the reflected light waves, the transmission of light through the medium is reduced significantly because of the conservation of energy. The transmission is the complement of reflection. Constructive interference in the reflection means destructive interference in the transmission. Such a change in the transmission of light can be put to good use in photonic devices (e.g., Q-switching, mode-locking, etc.).

The diffraction discussed above is usually called the Bragg diffraction and Eqs. (9.79) and (9.81) are the conditions of the Bragg diffraction.

(c) Diffraction by a traveling acoustic grating — Doppler shift in an isotropic medium If we consider the real situation in which a traveling acoustic wave is launched in the medium and we do not make any approximation, then the diffraction of the light wave will suffer a Doppler shift. That is to say, the observer in the laboratory reference frame will see that the diffracted wave's frequency will be shifted. Again, it is instructive to first review the elementary idea of the Doppler shift, which is then applied to the diffraction.

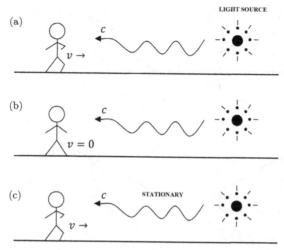

Fig. 9.5.   Schematic illustration of the Doppler effect.

(i) Classical picture of Doppler shift

As shown in Fig. 9.5(a), an observer moves at a velocity v towards a light source emitting light waves that travel at a velocity c towards the observer. The situation can be "decomposed" into two situations, one in which the observer is at rest and another in which the light wave is at rest (Fig. 9.5(b) and (c), respectively). In Fig. 9.5(b), the observer at rest "sees" c/λ sections of waves (each of length λ) passing him in one second, i.e., he "sees" a light frequency $v = c/\lambda$ when he is at rest. In Fig. 9.5(c), the moving observer "sees" a stationary light wave, and he passes v/λ sections of waves (each of length λ) in one second.

Combining Fig. 9.5(b) and (c), the observer "sees" $\left(\frac{c}{\lambda} + \frac{v}{\lambda}\right)$ sections of waves passing him in one second (case of Fig. 9.5(a)), i.e., he "sees" a light frequency.

$$v' = \frac{c}{\lambda} + \frac{v}{\lambda},$$
$$= v + \frac{v}{\lambda}, \tag{9.83}$$

i.e., there is a frequency shift of:

$$\Delta v = v' - v = \frac{v}{\lambda}. \tag{9.84}$$

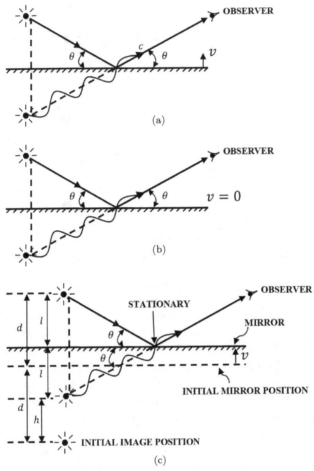

Fig. 9.6.   Doppler effect due to the motion of a mirror.

Equation (9.84) is the Doppler up-shift of the light's frequency measured by an observer moving towards the light source. If he moves away from the light source, v becomes negative and the frequency is Doppler down-shifted. In general, we can write the Doppler shift as either positive or negative:

$$\triangle v = \pm \frac{v}{\lambda}. \tag{9.85}$$

Now, in the case of diffraction by the acoustic wave, the observer "sees" the light waves reflected from the traveling acoustic index planes. We thus need to determine the frequency shift of one of the reflected "beams" from one of the acoustic index planes. This index plane can be considered as a "mirror", as shown in Fig. 9.6(a). The observer "sees" that the "mirror" moves in the vertical direction at a velocity v and the light wave moves from the image source towards the observer at a velocity c. Again, the situation can be "decomposed" into two, one with the "mirror" stationary, the other with the light wave stationary and the mirror moving, as shown in Fig. 9.6(b) and (c), respectively. In Fig. 9.6(b), the observer counts $\frac{c}{\lambda}$ sections of waves (each of length λ) in one second, i.e., he measures a frequency $v = \frac{c}{\lambda}$. In Fig. 9.6(c), the image moves at twice the speed of the mirror (in the non-relativistic regime). This can be seen from Fig. 9.6(c):

$$h + 2l = 2d,$$
$$h = 2(d - l). \tag{9.86}$$

Assume h is the displacement of the image in one second. Then v is the displacement of the "mirror" in one second. From the figure, $d = l + v$ or $d - l = v$. Substituting in Eq. (9.86), we have:

$$h = 2v, \tag{9.87}$$

i.e., the image speed is twice the "mirror" speed. To the observer, the image moves towards him at a speed of

$$v' = h \cos(90° - \theta),$$
$$= 2v \sin \theta. \tag{9.88}$$

This is equivalent to the observer moving towards a stationary light source at the speed v'. Thus, the situation in Fig. 9.6(c) is such that the observer passes $\frac{v'}{\lambda}$ sections of waves. Combining Fig. 9.6(b) and (c) gives the total number of sections of waves measured by the observer:

$$v' = \frac{c}{\lambda} + \frac{v'}{\lambda},$$
$$= v + \frac{2v \sin \theta}{\lambda}. \tag{9.89}$$

If the acoustic wave is propagating in the opposite direction (towards the bottom), v is negative. Thus the general shift of the frequency of the diffracted light is:

$$\triangle v = v' - v = \pm \frac{2v \sin \theta}{\lambda},$$

$$= \pm \frac{m}{D} v, \quad \text{(from Eq. (9.77))}$$

or

$$2\pi \triangle v = \pm m \frac{2\pi}{D} v,$$

$$\text{i.e.,} \quad \triangle \omega = \pm m K v. \tag{9.90}$$

The acoustic wave velocity is related to K and Ω of the acoustic wave by:

$$v = \frac{\Omega}{K}, \quad \text{(see Chap. I).} \tag{9.91}$$

Equation (9.90) thus becomes:

$$\triangle \omega = \pm m \Omega (m = 1, 2, 3, \ldots). \tag{9.92}$$

(ii) Quantum Picture of the Doppler shift

We can also describe the Doppler shift by using the quantum picture. Here, we assume that both the light wave and the acoustic wave are quantized so that we can describe them as photons and phonons, respectively. Such quantizations can be found in advanced books on quantum optics and solid-state physics. No details will be given here. The reader is thus asked to believe that the EM wave consists of a stream of photons each of energy $\hbar \omega$ and momentum $\hbar k$ while the acoustic wave consists of a stream of phonons each of energy $\hbar \Omega$ and momentum $\hbar k$. (k is in the direction of v, the acoustic wave velocity.) The diffracted light wave consists of a stream of photons each of energy $\hbar \omega'$ and momentum $\hbar k'$. Conservation of energy and

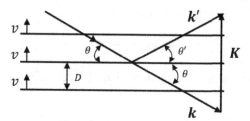

Fig. 9.7.   Quantum picture illustrating the Doppler effect in the diffraction of an EM wave by a traveling acoustic plane wave.

momentum requires that

$$\hbar\omega' = \hbar\omega + \hbar\Omega, \quad \text{(energy)}, \tag{9.93}$$

$$\hbar\mathbf{k}' = \hbar\mathbf{k} + \hbar\mathbf{K}, \quad \text{(momentum)}. \tag{9.94}$$

Figure (9.7) shows the relation (9.94), which becomes

$$\mathbf{k}' = \mathbf{k} + \mathbf{K}, \tag{9.95}$$

while Eq. (9.93) becomes

$$\omega' - \omega = \Omega$$

or

$$\triangle\omega = \Omega. \tag{9.96}$$

If the acoustic wave travels in the $-\mathbf{v}$ direction, $\mathbf{k}$ in Fig. 9.7 will reverse direction also and we have (from Fig. 9.7, with $\mathbf{k}$ reversed in direction):

$$\mathbf{k} = \mathbf{K} + \mathbf{k}' \tag{9.97}$$

and we have to write the conservation of energy as:

$$\hbar\omega = \hbar\Omega + \hbar\omega'$$

or

$$\omega' - \omega = -\Omega. \tag{9.98}$$

Combining Eqs. (9.96) and (9.98), we have:

$$\triangle\omega = \pm\Omega. \tag{9.99}$$

If the acoustic wave vector $|\mathbf{K}|$ is replaced by mK, it means we are dealing with the acoustic harmonics. Hence, the phonon energy is $m\hbar\Omega$ and Eq. (9.99) becomes:

$$\triangle\omega = \pm m\Omega, \qquad (9.100)$$

and this equation is identical to Eq. (9.92). Equation (9.94) and Fig. 9.7 give

$$k'\sin(\theta') + k\sin\theta = K. \qquad (9.101)$$

In general, $k' \neq k$, $\theta' \neq \theta$. But for very small change, we can approximate $k' = k$, $\theta' = \theta$, and Eq. (9.101) becomes:

$$2k\sin\theta = K, \quad (\text{no Doppler shift approximation}), \qquad (9.102)$$

which is identical to Eq. (9.81).

9.4 Diffraction by an Acoustic Wave in an Anisotropic Medium

If the medium is anisotropic, the index of refraction of the medium "seen" by the light beam is a function of the light propagation direction (Chapter 6). Since $\boldsymbol{k}$ and $\boldsymbol{k}'$ are not in the same direction, each will see a different index. It means that the condition for the Bragg diffraction (Eqs. (9.79) to (9.81)) is not $\theta = \theta'$ but $\theta \neq \theta'$ (Fig. 9.7).

We first analyze it qualitatively. Assuming that the polarization of the incident wave is linear and is oriented parallel to the polarization of the extraordinary wave inside the anisotropic medium. The diffracted wave, traveling in another direction $\hat{k}'$, will "see" two indices or two values of k', as shown in Fig. 9.8(a), in which we show the surface of wave normals of a uniaxial crystal. The lengths of $\boldsymbol{k}$ and the two possible values of k' are shown. However, for a fixed acoustic wave vector $\boldsymbol{K}$, only one value of k' will satisfy the Bragg diffraction condition. This is shown in Fig. 9.8(b) and (c). Either of these two conditions can be fulfilled separately but not simultaneously. This can be achieved by changing $\boldsymbol{K}$, the acoustic wave vector, i.e., changing its direction and frequency. Note: The law of reflection (Chapter 3) requires that $\boldsymbol{k}$ and $\boldsymbol{k}'$ lie in the same plane.

Quantitatively, let us assume that we have set the acoustic $\boldsymbol{K}$ such that Fig. 9.8(b) is valid. It is redrawn in Fig. 9.9. From trigonometric

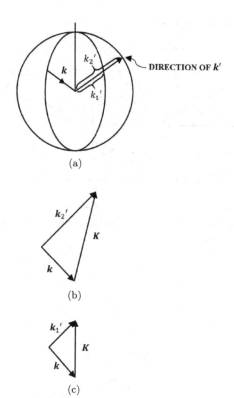

(a)

(b)

(c)

Fig. 9.8. Diffraction of light by an acoustic wave in an anisotropic medium (uni-axial in the figure).

relations, we see from Fig. 9.9 that

$$\begin{cases} k'^2 = k^2 + K^2 - 2kK \cos \alpha, \\ k^2 = k'^2 + K^2 - 2k'K \cos \beta', \end{cases}$$

Since,

$$\alpha = \pi/2 - \theta,$$
$$\beta = \pi/2 - \theta',$$

and the two relations become, after dividing them by K:

$$2k \sin \theta = \frac{k^2 - k'^2}{K} + K, \tag{9.103}$$

$$2k' \sin \theta' = \frac{k'^2 - k^2}{K} + K, \tag{9.104}$$

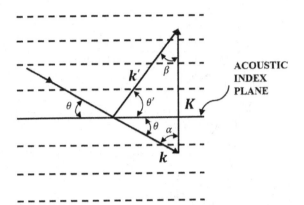

Fig. 9.9. Relationship between the incident and the acoustically diffracted angles θ and θ' in an anisotropic medium.

using

$$K = \frac{2\pi}{D}, \tag{9.105}$$

$$k' = \frac{n'\omega'}{c}, \tag{9.106}$$

$$k = \frac{n\omega}{c} \tag{9.107}$$

and assuming $\omega' \approx \omega$ because the Doppler shift is very small ($\omega' - \omega = \Omega \ll \omega$), which gives:

$$k' \approx \frac{n'\omega}{c}. \tag{9.108}$$

Equations (9.103) and (9.104) became:

$$2D\sin\theta = \frac{\lambda_o}{n} - \frac{D}{n\lambda_o}(n'^2 - n^2), \tag{9.109}$$

$$2D\sin\theta' = \frac{\lambda_o}{n'} - \frac{D}{n'\lambda_o}(n'^2 - n^2), \tag{9.110}$$

where λ_o is now the wavelength of light in a vacuum, D is the acoustic wavelength, and n and n' are the refractive indices "seen" by the incident and diffracted waves, respectively. Equations (9.109) and

(9.110) are the general conditions for the Bragg diffraction of light by an acoustic index wave in an anisotropic medium. The reason why they are general can be seen by referring to Fig. 9.8(a). Although we have assumed only one value of the incident k, the general case of two values of incident k is still the same for only one of the two values of k' to satisfy the momentum conservation condition (Fig. 9.8(b) or (c)), thus leading to Eqs. (9.109) and (9.110). The crystal can be biaxial instead of uniaxial and all the above arguments still apply, meaning that Eqs. (9.109) and (9.110) are indeed general conditions for the Bragg diffraction in an anisotropic medium. If there are acoustic harmonics present, we can write Eq. (9.105) as:

$$K \to K' \equiv Km, \quad m = 1, 2, 3, \ldots$$
$$\equiv \frac{2\pi}{D}m,$$
$$= \frac{2\pi}{\left(\frac{D}{m}\right)},$$
$$= \frac{2\pi}{D_m}, \tag{9.111}$$

where

$$D_m \equiv D/m. \tag{9.112}$$

Substituting into Eqs. (9.109) and (9.110), we have

$$2D_m \sin\theta = \frac{\lambda_o}{n} - \frac{D_m}{n\lambda_o}(n'^2 - n^2), \tag{9.113}$$

$$2D_m \sin\theta' = \frac{\lambda_o}{n'} - \frac{D_m}{n'\lambda_o}(n'^2 - n^2), \tag{9.114}$$

where D_m is given by Eq. (9.112). Both Eqs. (9.113) and (9.114) reduce to Eq. (9.79) in the special case of an isotropic medium, i.e., $n' = n$, $\theta' \simeq \theta$.

Under this condition, they become

$$2D_m \sin\theta = \frac{\lambda_o}{n}$$

or

$$2\frac{D}{m}\sin\theta = \lambda, \quad (\text{using Eqs. (9.82) and (9.112)})$$

or

$$2D\sin\theta = m\lambda.$$

Equations (9.109) and (9.110) or Eqs. (9.113) and (9.114) determine the incident and diffracted angles θ and θ' defined in Fig. 9.9. Not all values of λ and D will satisfy the real situation. In other words, there is a range of validity of λ and D, as illustrated by the following example.

Example: Let the medium be uniaxial. The incident wave sees an index n_e (i.e., its polarization is parallel to that of the extraordinary wave in the uniaxial medium). The diffracted wave sees the ordinary index n_o. Also, $n_o > n_e$ (negative uniaxial). We like to find a range of validity for the Bragg diffraction. Consider only the case $m = 1$, i.e., Eqs. (9.109) and (9.110), which can be re-written as

$$\sin\theta = \frac{1}{2n_e}\left\{\frac{\lambda_o}{D} - \frac{D}{\lambda_o}(n_o^2 - n_e^2)\right\}, \tag{9.115}$$

$$\sin\theta' = \frac{1}{2n_o}\left\{\frac{\lambda_o}{D} + \frac{D}{\lambda_o}(n_o^2 - n_e^2)\right\}. \tag{9.116}$$

These conditions are valid only if θ and θ' are real, or

$$-1 \leq \left\{\begin{array}{c}\sin\theta \\ \sin\theta'\end{array}\right\} \leq 1. \tag{9.117}$$

Using Eqs. (9.115) and (9.116), Eqs. (9.117) becomes

$$-1 \leq \frac{D}{\lambda_o}\cdot\frac{1}{2n_e}\left\{\left(\frac{\lambda_o}{D}\right)^2 - (n_o^2 - n_e^2)\right\} \leq 1, \tag{9.118}$$

$$-1 \leq \frac{D}{\lambda_o}\cdot\frac{1}{2n_o}\left\{\left(\frac{\lambda_o}{D}\right)^2 + (n_o^2 - n_e^2)\right\} \leq 1. \tag{9.119}$$

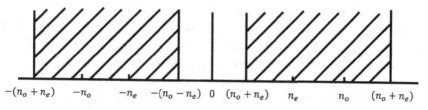

RANGE OF VALIDITY: SHADED AREA

Fig. 9.10.   Range of validity (shaded region) during the diffraction of light by an acoustic wave in a uniaxial medium.

Let

$$q \equiv \frac{\lambda_o}{D}, \tag{9.120}$$

Eq. (9.118) becomes two inequalities:

$$q^2 - 2n_e q - (n_o^2 - n_e^2) \leq 0, \tag{9.121}$$

$$q^2 + 2n_e q - (n_o^2 - n_e^2) \geq 0, \tag{9.122}$$

while Eq. (9.119) becomes:

$$q^2 - 2n_o q + (n_o^2 - n_e^2) \leq 0, \tag{9.123}$$

$$q^2 + 2n_o q + (n_o^2 - n_e^2) \geq 0, \tag{9.123'}$$

$$\text{for} \quad (n_o > n_e).$$

It is now left as an <u>exercise</u> for the reader to obtain the following condition of validity using Eqs. (9.121) to (9.113).

$$|n_o - n_e| \leq \frac{\lambda_o}{D} \leq |n_o + n_e|. \tag{9.124}$$

This is shown in Fig. 9.10 (shaded region).

9.5   Higher Order Diffraction by an Acoustic Wave

So far, in the previous two sections, we consider only the lowest order diffraction of infinitely large plane light waves by infinitely large plane acoustic waves, namely, $m = 1$ in Eq. (9.79) or (9.80). Higher-order diffraction $m \geq 2$ will take place under any one of the following conditions. We shall analyze them only qualitatively.

(i) The interaction between the light wave and the acoustic wave is limited in a very small region. This implies that even if the total acoustic wave is still a plane wave across a large region of the medium, the scattered light wave cannot be plane anymore because the small interaction region cannot re-radiate plane waves. By Huygen's principle, each tiny element in the interaction region radiates spherical waves that will re-combine with one another, resulting in higher-order constructive interference at a distance away. Think of this as similar to the diffraction of a plane light wave by a small aperture. The phase of the optical wave is modified, giving rise to higher-order patterns (fringes) in the far field. Thus, the acoustic wave in the small region modulates only the phase of the light wave but not the amplitude; hence, *phase grating*. Such a situation can be achieved by limiting the transmission of the light wave into the interaction region by an aperture very close to the diffraction region.

(ii) If the frequency of the acoustic wave is very low so that the acoustic wavelength D is very large, then, even if both the optical and acoustic waves are plane waves, Eqs. (9.79) and (9.80) are also valid at values of m larger or equal to 2.

(iii) If the acoustic wave is not a plane wave, even if the light wave is plane and the interaction region is large, higher-order diffraction occurs in the following sense. Referring to Fig. 9.11(a), we assume that the acoustic index wavefronts are curved surfaces. These surfaces are equivalent to the combination of a continuous set of plane waves that are tangent to the curved surfaces. A plane light wave incident onto these curved acoustic index waves will suffer multiple diffraction (scattering) by successive plane acoustic waves (Fig. 9.11(b)). This is equivalent to the wave vector picture of multiple photon-phonon scattering (Fig. 9.11(c)), in which every wave vector (k or K) represents a plane wave. The successive change in the directions of K_1, K_2, K_3, etc., represents the curvature of the acoustic wavefront. Comparing Figs. 9.11(c) and 9.7, and using Eq. (9.93), we see that (assuming that the frequency of all K's is Ω_0):

$$\omega_1 = \omega_0 + \Omega_0,$$

$$\omega_2 = \omega_1 + \Omega_0 = \omega_0 + 2\Omega_0,$$

$$\omega_{m+1} = \omega_0 + m\Omega_0.$$

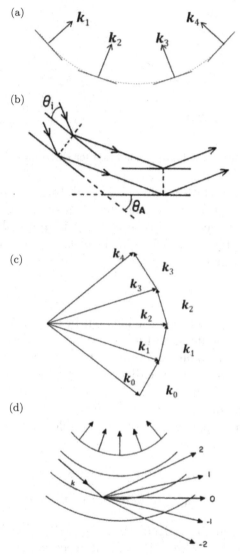

Fig. 9.11. Higher order acoustic diffraction.

That is to say, the multiple diffraction gives rise to much high-order constructive interference. In general, one speaks of orders ± 1, ± 2, ± 3, $\ldots \pm m \ldots$, as shown in Fig. 9.11(d).

All the above three cases that generate high-order diffractions are sometimes called "*Raman-Nath*" *diffraction*. One physical way to

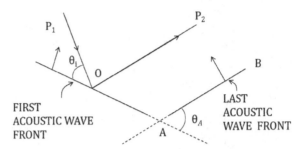

Fig. 9.12.   Illustrating the limiting condition for the Raman-Nath diffraction.

distinguish Raman-Nath (or high-order) diffraction from the lowest order Bragg diffraction is the following. Consider Fig. 9.12.

The limit at which multiple scattering is impossible is that the reflected beam OP_2 is parallel to the last plane acoustic wave surface AB. This means that

$$\theta_i = \theta_A,$$

where θ_i is the angle between the incident beam and the first acoustic index plane it encounters, and θ_A is the angle between the first and last acoustic index plane (assuming a concave acoustic surface, see Fig. 9.11). θ_A is essentially the divergence (convergence) angle of the acoustic wavefront. Thus, when the incident angle θ_i (as defined in Fig. 9.12) is equal or greater than the divergence angle of the acoustic wavefront, no multiple diffraction is possible. Hence, the following criteria are physically sound:

$$\theta_i \geq \theta_A, \quad \text{Bragg diffraction,}$$

$$\theta_i < \theta_A, \quad \text{Raman-Nath diffraction.}$$

However, as can be judged from Fig. 9.12, if θ_i is only *slightly smaller* than θ_A, the beam OP_2 will intersect the last acoustic plane AB at a very far away point. This, in practice, is impossible. Thus, for practical purposes, $\theta_i < \theta_A$ is not sufficient.

We can analyze this problem more closely by looking at a specific example. Assuming that the acoustic wave has a Gaussian spherical converging wavefront similar to an optical Gaussian spherical wave. Using the same concept for defining the angle of divergence of an optical spherical wave in the far field (Chapter 5, Eq. (5.37)), the

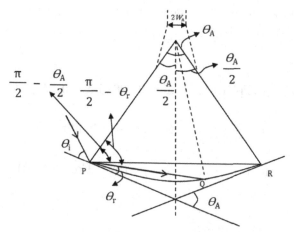

Fig. 9.13.   Geometrical analysis of acoustic diffraction of various orders.

full angle of divergence of the acoustic wave in the far field is

$$\theta_A \cong \frac{2D}{\pi w_A}, \tag{9.125}$$

where D is the acoustic wavelength, and $2w_A$ is the acoustic beam waist (diameter). As shown in Fig. 9.13, the spherical acoustic index wavefront is represented by the curve PQR. That is to say, the acoustic wave has a limited extent. Assume the extreme case in which a light beam is incident at P. From the figure, the reflected beam PQ hits the acoustic wavefront again at point Q. This is the condition for Raman-Nath multiple scattering. The limit for multiple scattering is when the reflected beam hits the other extremity of the acoustic wave, point R. We can thus say that the condition for Raman-Nath multiple diffraction is (see Fig. 9.13):

$$\frac{\pi}{2} - \theta_r \geq \frac{\pi}{2} - \theta_A/2, \quad \text{(Raman-Nath diffraction)}. \tag{9.126}$$

The equal sign corresponds to when the reflected beam hits point R. Since $\theta_r = \theta_i$ (law of reflection), Eq. (9.126) becomes

$$\frac{\theta_A}{2} \geq \theta_i, \quad \text{(Raman-Nath diffraction)}. \tag{9.127}$$

The incident angle θ_i has to satisfy the lowest order Bragg diffraction condition (Eq. (9.79), $m = 1$)

$$2D\sin\theta_i = \lambda. \tag{9.128}$$

Substituting Eqs. (9.128) and (9.126) into (9.127), we obtain:

$$\frac{D}{\pi w_A} \geq \sin^{-1}\left(\frac{\lambda}{2D}\right), \quad \text{(Raman-Nath diffraction)}. \tag{9.129}$$

If θ_i is very small, $\theta_i - \sin\theta_i = \frac{\lambda}{2D}$, (from Eq. (9.128)), Eq. (9.129) becomes

$$\frac{D}{\pi w_A} \geq \frac{\lambda}{2D},$$

$$\frac{2D^2}{\pi \lambda w_A} \geq 1, \quad \text{(Raman-Nath diffraction)}. \tag{9.130}$$

In summary, the conditions for the Bragg and Raman-Nath diffractions by Gaussian spherical acoustic waves of beam waist W_A and acoustic wavelength D are From Eq. (9.127)

$$\theta_i \begin{cases} > \theta_A/2 & \text{Bragg} \\ \leq \theta_A/2 & \text{Raman-Nath} \end{cases}$$

Or from Eq. (9.129)

$$\sin^{-1}\left(\frac{\lambda}{2D}\right) \begin{cases} > \dfrac{D}{\pi w_A} & \text{Bragg} \\ \leq \dfrac{D}{\pi w_A} & \text{Raman-Nath} \end{cases}$$

For small θ_i, from Eq. (9.130)

$$\frac{2D^2}{\pi \lambda w_A} \begin{cases} < 1 & \text{Bragg} \\ \geq 1 & \text{Raman-Nath} \end{cases}$$

9.6   Closing Remarks

While the electro-optic effect is the consequence of electric field-induced anisotropic polarization in a medium, the acousto-optic effect is the consequence of mechanical (acoustic) force-induced elastic anisotropy. In both cases, one should pay attention to the polarization state of the light (laser) wave that passes through the medium because of the induced anisotropy. The electro-optic medium is essentially an electric field-induced wave plate, while the acousto-optic medium can be considered as a mechanical (acoustic) force-induced grating. In the latter case, light wave diffracted by the acoustic index grating can be used for laser beam scanning and modulation, mode locking, Q-switching, spectrum analysis, etc. For example, in the case of Q-switching, the diffraction of laser radiation in an acousto-optic medium inside the laser cavity acts as a loss mechanism, thus prohibiting laser oscillation. This allows the Q of the cavity to build up (see Chapter 8). When the acoustic field is turned off after a while, oscillation starts, and a giant laser pulse is generated.

No attempt is made in this chapter to go into the detail of any application of the acousto-optic effect. The reader is referred to sources such as (Yariv and Yeh, 1984).

Reference

A. Yariv and P. Yeh (1984). Optical Waves in Crystals: Propagation and Control of Laser Radiation. Wiley-Intersciences, New York.

Chapter 10

Magnetic Field-Induced Anisotropy

When a mechanical force and an electric field can both induce optical anisotropy in a medium, it is natural to ask if a magnetic field will also induce anisotropy. The answer is "yes". The magnetic field can be static or the magnetic component of the incident electromagnetic (EM) field. Both give rise to a change in some special media, resulting in a rotation of the electric field vector (polarization state) of the EM wave that passes through them. This short chapter explains qualitatively the physical phenomena underlying the magnetic field-induced anisotropy. There are two types of magnetic field-induced anisotropy; one is natural and the other is "artificial". The former is traditionally called *optical activity* and the latter is called the *Faraday rotation*. At the end of the chapter, three examples of application will be discussed. They are the *Faraday rotator*, *Faraday isolator*, and *magneto-optical switch*.

10.1 Optical Activity

The phenomenon occurs in any natural material that does not have inversion symmetry, while the molecular structure in the material has some kind of spiral nature. Thus, apart from the normal dielectric response of the material to the electric component of an EM wave propagating in it, the spiral structure of the molecules will also respond to the magnetic component of the EM wave. This is because some non-bonding electrons on such spiral molecules will move around the spiral when there is a time-varying magnetic field

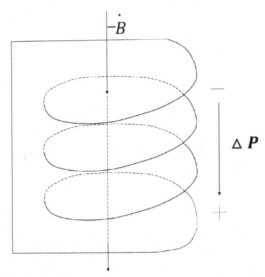

Fig. 10.1. Similarity of response of a spiral molecule to a solenoid when placed in a time-varying magnetic field.

passing through it. This is similar to a solenoid (or coil) put in a time-varying magnetic field (Fig. 10.1). A current will flow through the solenoid according to Faraday's law:

$$-\frac{\partial \boldsymbol{B}}{\partial t} = \nabla \times \boldsymbol{E}. \tag{10.1}$$

Because the magnetic field in an EM wave varies rapidly, the accumulation of charges (+ and −) at the top and bottom of the solenoid (Fig. 10.1) also rapidly changes sign. Thus, there is a time-varying-induced dipole moment. If we consider that the medium consists of a distribution of many molecular solenoids, then for every dielectric response of the medium in a local small volume, there is also a magnetic response that gives rise to a change in polarization, $\boldsymbol{\Delta P}$. From Eq. (10.1) and Fig. 10.1, we see that

$$\nabla \boldsymbol{P} \propto \left(-\frac{\partial \boldsymbol{B}}{\partial t}\right) = \nabla \times \boldsymbol{E}. \tag{10.2}$$

We assume a plane EM wave propagating through the medium (cf. Eqs. (6.26) and (6.27)), so that

$$\nabla = -i\boldsymbol{k}.$$

Then Eq. (10.2) becomes:

$$\Delta \boldsymbol{P} \propto -i\boldsymbol{k} \times \boldsymbol{E},$$

$$\Delta \boldsymbol{P} = -i\alpha \boldsymbol{k} \times \boldsymbol{E}, \tag{10.3}$$

where α is the proportionality constant. If we define

$$\boldsymbol{G} \equiv -\frac{\alpha \boldsymbol{k}}{\epsilon_0}, \tag{10.4}$$

Eq. (10.3) becomes

$$\Delta \boldsymbol{P} = +i\epsilon_0 \boldsymbol{G} \times \boldsymbol{E}. \tag{10.5}$$

The total field $\boldsymbol{D}$ in the medium is now

$$\boldsymbol{D} = \boldsymbol{P} + \epsilon_0 \boldsymbol{E} + \Delta \boldsymbol{P},$$

$$= \epsilon \boldsymbol{E} + \Delta \boldsymbol{P},$$

$$= \epsilon \boldsymbol{E} + i\epsilon_0 \boldsymbol{G} \times \boldsymbol{E}. \tag{10.6}$$

Here, ϵ is the dielectric tensor of the medium.

10.2 Faraday Rotation

Consider the propagation of a plane EM wave in a medium. The electric field of the wave sets the electrons in the path of the EM wave in motion. Let the instantaneous velocity of the electron be v. Thus:

$$m\dot{v} = e\boldsymbol{E} \propto e^{i\omega t},$$

$$\boldsymbol{v} \propto \int e^{i\omega t} dt = \frac{1}{i\omega} e^{i\omega t} \propto -i\boldsymbol{E}. \tag{10.7}$$

If we now apply an external constant magnetic field $\boldsymbol{B}$ along the propagation direction $\hat{k}$ of the EM wave, the moving electron will experience a magnetic force due to the external field $\boldsymbol{B}$, and is given by the Lorentz force:

$$\boldsymbol{F} = e\boldsymbol{v} \times \boldsymbol{B}. \tag{10.8}$$

The electronic charge e is negative. Such a force will have the effect of generating an additional separation of positive and negative charge,

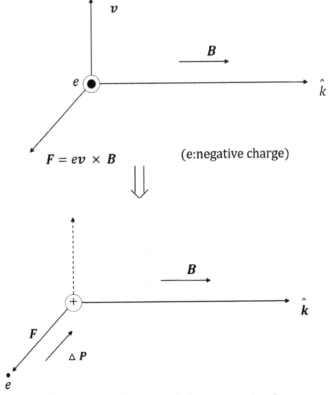

Fig. 10.2. Fundamental consideration of electron motion in a magnetic field leading to the Faraday rotation.

as shown in Fig. 10.2. There is then an additional polarization in a local volume:

$$\Delta P \propto -F = -ev \times B, \tag{10.9}$$

which becomes, using Eq. (10.7):

$$\Delta P \propto -e(-i\mathbf{E}) \times B,$$

or

$$\Delta P = iae\mathbf{E} \times B, \quad \text{a : proportionality constant,}$$
$$= ia(-1)|e|(-B \times E),$$
$$= ia|e|B \times E. \tag{10.10}$$

Let

$$\epsilon_0 \gamma \boldsymbol{G} \to \equiv a|e|\boldsymbol{B}, \tag{10.11}$$

where $\gamma = \underline{\text{magnetogyration coefficient}}$.

Equation (10.10) becomes

$$\Delta \boldsymbol{P} = i\gamma\epsilon_0 \boldsymbol{G} \times \boldsymbol{E}. \tag{10.12}$$

The total field $\boldsymbol{D}$ induced in the medium is then

$$\boldsymbol{D} = \epsilon\boldsymbol{E} + \Delta\boldsymbol{P} = \epsilon\boldsymbol{E} + i\epsilon_0\gamma\boldsymbol{G} \times \boldsymbol{E}. \tag{10.13}$$

10.3   Polarization Rotation Due to Magnetic Field-Induced Anisotropy

Equations (10.6) and (10.12) both show that the field $\boldsymbol{D}$ is equal to the normal reaction $\varepsilon\boldsymbol{E}$ plus a complex term. This complex term is due to the motion of the charges caused by the magnetic field. It is well known in electromagnetism that no work is done when charges are deviated by magnetic fields. This is also physically true in the above two cases and can be quantitatively seen as follows. The energy in the EM field is proportional to $(\boldsymbol{D}\cdot\boldsymbol{E})$. From both Eqs. (10.6) and (10.13), the complex term's contribution is $(\boldsymbol{G} \times \boldsymbol{E}) \cdot \boldsymbol{E} = 0$, because $\boldsymbol{G} \times \boldsymbol{E} \perp \boldsymbol{E}$.

Thus, the magnetic term $(\boldsymbol{G} \times \boldsymbol{E})$ does not contribute to any energy dissipation as in the case of the complex dielectric constant of metal optics. ($\boldsymbol{G}$ is called the *Gyration Vector.*)

One can now use the new expressions of $\boldsymbol{D}$ (Eqs. (10.6) and (10.13)) and go through the same calculation as in Section 6.2, obtaining Fresnel's equation of wave normals and the ray surfaces. The results show that the complex term $\pm i\boldsymbol{G}\times\boldsymbol{E}$ has the effect of transforming an incident linearly polarized EM wave into two circularly polarized waves rotating in opposite directions. These two waves of counter-rotating circular polarizations "see" two different indices of refraction in the medium. Let them be n_1 and n_2. Thus, when the waves pass through the medium, say, in the form of a plate of thickness l, there is a phase difference between the two waves given by:

$$\Delta\epsilon = (k_1 - k_2)l = \frac{\omega}{c}(n_1 - n_2)l. \tag{10.14}$$

At the exit, the two circularly polarized waves recombine to form a linearly polarized wave again, except that the latter has been rotated with respect to the incident linear polarization by an angle θ where

$$\theta = \frac{l}{2}(k_1 - k_2) = \frac{l\omega}{2c}(n_1 - n_2). \tag{10.15}$$

In the case of the *Faraday rotation*, θ is proportional to the external magnetic field B and the thickness l and we have:

$$\theta \propto Bl$$

or

$$\theta = \frac{V}{\mu}Bl, \tag{10.16}$$

where we have set the proportionality constant as V/μ, with μ the magnetic permeability of the medium. V is called the *Verdet constant*. We assume an observer looks into the laser beam. A positive Verdet constant corresponds to anticlockwise rotation, when the direction of propagation is parallel to the magnetic field and to clockwise rotation when the direction of propagation is anti-parallel. In this sense, if a light beam passes through a magnetic material in the magnetic field and is reflected back through it, the rotation doubles.

Combining Eqs. (10.15) and (10.16):

$$\frac{V}{\mu}Bl = \frac{l\omega}{2c}(n_1 - n_2),$$

$$V = \frac{\mu\omega}{2cB}(n_1 - n_2). \tag{10.17}$$

There are many examples of optically active materials, such as liquid crystals, quartz, cane sugar solution, etc. In the case of quartz, which is also a uniaxial crystal, one might ask if the optical activity will influence the optical anisotropy of the medium. In principle, yes. But the effect is negligible if the wave is traveling in a direction significantly different from that of the optic axis. When the wave travels in or near the optic axis direction, the optical activity becomes more important. But then the anisotropy becomes isotropic or nearly so because the wave propagates in or near the direction of the optic axis.

Optical activity differs in practice from the Faraday rotation. In the former case, the direction of rotation of linear polarization of an EM wave propagating through the medium is the same when an observer looks into the wave, regardless of whether the wave propagates in one or the opposite direction. Hence, if a linearly polarized wave passes through an optically active medium and is reflected back along the same path, the net rotation is null. In the case of the Faraday rotation, a back reflection of the linearly polarized light wave (now in the opposite direction to the externally applied magnetic field) will double the angle of rotation. Such a Faraday rotation is used in laser amplifier chains to isolate reflected pulses, and the device is called a *Faraday rotator*.

10.4   Magneto-optic Instruments

The previous sections qualitatively explain the physical phenomena underlying the magnetic field-induced anisotropy. It is known that an "artificial" magnetic field induces a *Faraday rotation*. Nowadays, the operation of many magneto-optic instruments is based on a Faraday rotation. This section introduces three typical magneto-optic instruments: *Faraday rotator, Faraday isolator, and magneto-optical switch. Faraday rotators and isolators* are widely used inside commercial pulsed laser systems such as solid laser oscillators and amplifiers, while *magneto-optical switches* are used in laser field modulation.

10.4.1   Faraday rotator

A *Faraday rotator* is a magneto-optic device in which light is transmitted through a transparent medium exposed to a magnetic field. The magnetic field lines have approximately the same direction as the optical beam direction, or the opposite direction. If the light is initially linearly polarized, its polarization direction can be continuously rotated during the passage through the medium. According to Eq. (10.16), the Faraday rotation angle θ is proportional to the external magnetic field B and the thickness l of the device.

For a *Faraday rotator*, the change in polarization direction is defined only by the magnetic field direction and the sign of the *Verdet constant*. If a linearly polarized beam is sent through a *Faraday rotator* and back again after reflection from a mirror, the polarization

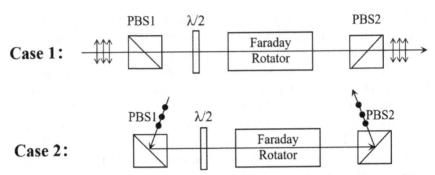

Fig. 10.3.   Two typical "Faraday rotator+wave plate+polarization beam splitter (PBS)" schemes. The double arrows indicate horizontally polarized beams, while dots represent vertically polarized beams. Both PBS1 and PBS2 can transmit the horizontally polarized light, and reflect the vertically polarized one.

changes of the two passes add up. This non-reciprocal behavior distinguishes Faraday rotators from the arrangement of wave plates and polarizers.

Figure 10.3 indicates two typical "Faraday rotator + wave plate + polarization beam splitter (PBS)" schemes.

In Case 1, a horizontally polarized beam is directed through the PBS1 (see Fig. 10.3) from the left to the right. The half-wave plate and the *Faraday rotator* are designed so that they will each rotate the polarization of the beam by 45° but in the opposite directions. Let us say the magnetic field in this case is directed in the same direction of the laser propagation. In this case, the output beam is still horizontally polarized, which can pass through the PBS2. On the other hand, if the horizontally polarized beam goes back (is reflected back) from the right to the left along the same axis, PBS2 will transmit the beam totally while the *Faraday rotator* and the half wave plate will each rotate the polarization by 45° but now in the same direction. This is because when a light is passed through a Faraday rotator and reflected back through it, the rotation doubles (see explanation after Eq. (10.6)). However, if it is a half-wave plate instead of the Faraday rotator, the polarization will not change. The polarization rotation angles (two 45°) will add up, which finally leads to 90° rotation of the polarization. Thus, the output from the half-wave plate becomes vertically polarized and is blocked by the PBS1. Case 1 has been applied in *Faraday isolators*. It will be introduced at the end of this section.

In Case 2, the reflection from the PBS is directed to the laser cavity (see Fig. 10.3). If the beam goes from left to right, the vertically polarized input beam is reflected by the PBS1. The half-wave plate and the *Faraday rotator* will rotate the light by 45° in the opposite direction. In this sense, the output beam is still vertically polarized, which can be reflected into the laser cavity by PBS2. On the other hand, if the beam goes back from the right to the left along the same propagation axis, the polarization rotation angle by the Faraday rotator and the half wave plate will add up as in case 1, which finally leads to 90° rotation of the polarization. Thus, the output horizontally polarized light is transmitted by the PBS1 and cannot enter the cavity.

10.4.2 Faraday isolators

A *Faraday isolator* is composed of a Faraday rotator, a half-wave plate, and two polarizers (see Fig. 10.4). In many applications, the Faraday rotator rotates the polarization direction of the incident light by a fixed 45°. Figure 10.4 indicates the evolution of the polarization states of the light when it goes through the isolator. Here, we call PBS 1 the "V-Polarizer", indicating that the transmitted light is polarized vertically (90°). And PBS 2 is named "H-Polarizer", indicating that the transmitted light is polarized horizontally (0°).

According to Case 1 of Fig. 10.4, when the incident light goes through an isolator in the forward direction, the Faraday rotator

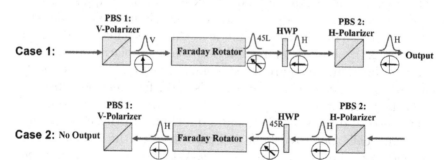

Fig. 10.4. Polarization evolution for a light beam during passage through a Faraday isolator in the forward direction (Case 1) and backward direction (Case 2).

Note: PBS: polarization beam splitter.

and the half-wave plate each rotates the polarization by 45°, successively. After that, the light is horizontally polarized (0°) when it reaches PBS 2. In this case, it can go through PBS 2 without attenuation. In contrast, according to Case 2 of Fig. 10.4, if the light is reflected back along the same propagation axis and goes through the Faraday rotator again, the polarization rotation by the half-wave plate and Faraday rotator will further rotate the polarization into the horizontal direction (0°) when it reaches PBS 1. Since horizontally polarized light cannot pass through PBS 1, it will be reflected out. Hence, it is called an isolator.

10.4.3 Magneto-optical switch

Magneto-optical switch uses Faraday's magneto-optical effect for optical switching. Ferromagnetic medium is the key element of a magneto-optical switch. Ferromagnetic materials, e.g., $(CO, Cr, Fe)_3O_4$, are usually used due to high specific Faraday rotations.

Magneto-optical switch has various structures. Figure 10.5 shows the schematic diagram for one of them. Here, the transmission axes for the V-Polarizer and H-Polarizer are orthogonal to each other. The magnetic field provided by a solenoid is applied to the transparent ferromagnetic medium. In this experiment, the ferromagnetic medium is transparent to the input beam, while the rod of the solenoid is a dielectric object. The ferromagnetic medium is isotropic without external magnetic fields. It can turn into a birefringent medium with an external magnetic field. A fixed current is applied to the solenoid so as to generate a constant magnetic field. Once the current is switched on, the ferromagnetic medium becomes a birefringent medium. A π phase delay results between the polarization components of the input beam projected along the fast and slow axes of the ferromagnetic medium. The optical axis of the ferromagnetic medium is specially orientated in order to rotate the input polarization by 90° (from vertical direction to horizontal direction). In this case, the initially vertically polarized light can pass through the H-Polarizer at the end of the *magneto-optical switch*. When no current is applied, the input polarization direction will not change since there is no magnetic field applied. Then the H-Polarizer will block the beam; thus, the path is switched off. In Fig. 10.5, a part of the reflection from

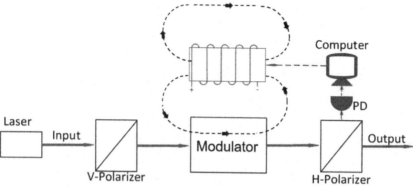

Fig. 10.5. Schematic diagram of a magneto-optical switch. PD: Photoelectric detector. V-Polarizer transmits vertically polarized light, while H-Polarizer transmits the horizontally polarized light. The magnetic field lines from the solenoid to the ferromagnetic medium are indicated by the black dashed line.

the H-Polarizer is taken by the monitoring terminal. It works as a feedback so as to stabilize the current on the solenoid.

A pulse slicer inside some high-power ultrafast laser system is actually a magneto-optical switch because of its fast switching time. Compared with the traditional mechanical optical switch, a magneto-optical switch has the advantages of short switching time, high stability, and easy to integrate. While compared with other non-mechanical optical switches, it also holds the advantages of low-driving voltage and small crosstalk.

Closing Remarks

Nowadays, optical instruments based on magnetic field induced anisotropy is rather popular, in particular, in mode-locked fiber lasers (oscillators and amplifiers). For instance, in a fiber oscillator, the unidirectional operation of the oscillating beam can be guaranteed by a *Faraday isolator* or a *Faraday rotator* with a half-wave plate and polarization beam splitter (PBS). Besides, in a cascade laser amplification system, *Faraday isolators* are implemented in between each stage of amplifiers or between the oscillator and the amplifier in order to prevent parasitic oscillation from back-reflected light.

Chapter 11

Importance of Anisotropy in Second-Harmonic Generation

Since the invention of the laser, nonlinear optical phenomena were studied extensively, resulting in the possibility of generating very intense coherent radiation from the infrared to the ultra-violet. The field is now so vast that there is no hope to cover everything, even qualitatively, in only one small chapter. Excellent texts exist in the market and the interested reader should consult any of these texts. The present chapter will introduce the subject and show the importance of optical anisotropy in nonlinear optics using second-order nonlinear processes as an example. Third-order or higher-order nonlinear processes will be only briefly introduced.

11.1 Introduction

In nonlinear optics as in all other electromagnetic phenomena, we normally ask what the material response is, i.e., what the induced polarization in the medium is. This is because the induced polarization is the source of the re-radiation by the medium described by the wave equation (assume a lossless medium):

$$\nabla^2 \boldsymbol{E} - \frac{1}{v^2}\frac{\partial^2 \boldsymbol{E}}{\partial t^2} = \mu_o \frac{\partial^2 \boldsymbol{P}}{\partial t^2}, \tag{11.1}$$

(cf. any text on electromagnetism) where $\boldsymbol{P}$ is the induced polarization in the medium and $\boldsymbol{E}$ is the electric field vector of the electromagnetic (EM) wave propagating in the medium.

The term on the right hand side of Eq. (11.1) is the source of the re-radiation. Normally, using ordinary weak light, $\boldsymbol{P}$ is linear in $\boldsymbol{E}$ (or the medium is linear):

$$\boldsymbol{P} = \epsilon_0 x \boldsymbol{E}, \tag{11.2}$$

even if y is a tensor, i.e., even if the medium is anisotropic. However, when an intense laser beam propagates in the medium, the medium becomes nonlinear. That is, the laser-induced polarization in the medium has the following general form (following (Yariv and Yeh, 1984)):

$$\boldsymbol{P} \equiv P_i, (i = 1, 2, 3),$$
$$= \epsilon_0 x_{ij} E_j + 2d_{ijk} E_j E_k + 4x_{ijkl} E_j E_k E_l + \cdots , \tag{11.3}$$

where repeated subscripts (j, j), (k, k) and (l, l) mean summation over $(1, 2, 3)$, the three Cartesian coordinates. $x_{ij}, d_{ijk}, x_{ijkl}$, etc., are the linear, second, third, etc., order susceptibilities, respectively. In principle, one substitutes the expression for $\boldsymbol{P}$ in Eq. (11.3) into the wave equation (11.1), solves it, and obtains the answer for the re-radiated field $\boldsymbol{E}$. The task is, of course, not trivial, and we shall not do any such calculation here. We just remark that each of the nonlinear terms in Eq. (11.3) is proportional to the product of two or more fields. These electric fields could come from the same laser beam (thus same frequency), and we talk about harmonic generation. They could come from different laser beams of the same or different frequencies, and we talk about sum and difference frequencies generation, parametric oscillation, etc. Phase-matching conditions relate the polarizations of different frequencies. For second-harmonic generation, only the second term on the right hand side of Eq. (11.3) is important and kept (see below).

11.2 Second-harmonic Generation

We now discuss the consequence of second-harmonic generation (SHG) in an anisotropic medium by skipping all the calculations under the following assumptions:

(1) The medium does not have an inverse symmetry (if so, the second-order coefficients d_{ijk} will be identically zero).
(2) The medium is lossless.

(3) Neglect all magnetic effect, i.e., $\mu \approx \mu_0$.

(4) Plane wave approximation, i.e., the solutions of all the relevant fields are plane waves propagating in the z-direction, i.e., the incident laser beam's electric field is expressed as:

$$
\begin{aligned}
E_i^{(\omega)}(z,t) &= E_i^{(\omega)}(z)\cos(\omega t - kz), \\
&= E_i^{(\omega)}(z)\,Re\{e^{i(\omega t - kz)}\}, \\
&= E_i^{(\omega)}(z)\{e^{i(\omega t - kz)} + e^{-i(\omega t - kz)}\}, \\
&\equiv \frac{1}{2}E_i^{(\omega)}(z)\{e^{i(\omega t - kz)} + c.c.\}.
\end{aligned}
\tag{11.4}
$$

where $c.c.$ is the complex conjugate. The subscript i defines the polarization direction of E, and the superscript denotes the frequency. Because the wave propagates in the z-direction, $i = 1$, 2 or x, y only. Similarly, the field of the second harmonic and the polarization of the second harmonic are respectively:

$$
E_k^{(2\omega)}(z,t) = \frac{1}{2}E_k^{(2\omega)}(z)\{e^{i(\omega t - kz)} + c.c.\},
\tag{11.5}
$$

$$
P_j^{(2\omega)}(zt) = \frac{1}{2}P_j^{(2\omega)}(z)\left\{\frac{1}{2}e^{i(2\omega t - kz)} + c.c.\right\}.
\tag{11.6}
$$

Substituting Eqs. (11.4) to (11.6) into the wave equation, with the dependence of $P_j^{(2\omega)}$ on the field (Eq. (11.3)) explicitly expressed, and keeping only the second term on the right hand side, one obtains an expression for the intensity $I_j^{(2\omega)}$ of the second harmonic radiation under an additional assumption of low conversion efficiency, i.e., $\frac{dE_i^{(\omega)}}{dz} \simeq 0$:

$$
I_j^{(2\omega)} = f(\omega, n, d)I_i^{(\omega)}I_\ell^{(\omega)}L^2\frac{\sin^2(\frac{1}{2}\Delta kL)}{(\frac{1}{2}\Delta kL)^2},
\tag{11.7}
$$

where $f(\omega, n, d)$ is a function of the frequency ω, index n is the second harmonic and second harmonic coefficient d. The subscripts j, I and l of the I's are the polarization directions of the radiation at the appropriate frequencies (2ω) or (ω) denoted by the superscripts. L is the interaction length in the medium, and Δk is given by:

$$\Delta k \equiv k_j^{(2\omega)} - k_i^{(\omega)} - k_l^{(\omega)}, \tag{11.8}$$

where $k_j^{(2\omega)}$ is the wave vector of the second harmonic wave polarized in the j-direction, and $k_i^{(\omega)}$ and $k_l^{(\omega)}$ are the wave vectors of the fundamental frequencies polarized in the i- and l-directions.

We analyze briefly Eq. (11.7), keeping in mind that we wish to generate the highest possible $I_j^{(2\omega)}$. That is to say, we want to maximize the expression for $I_j^{(2\omega)}$. We note the following:

(1) The explicit expression for $f(\omega, n, d)$ is:

$$f(\omega, n, d) = 8 \left(\frac{\mu_0}{\epsilon_0} \right)^{3/2} \frac{\omega^2}{n^3} d^2. \tag{11.9}$$

For a fixed frequency, we can choose materials of as large a d-value as possible while the index n of the second harmonic in the material cannot be changed.

(2) One can increase either $I_i^{(\omega)}$ or $I_l^{(\omega)}$ (or both) to increase $I_j^{(2\omega)}$. Note that if $i = l$, i.e., if the second harmonic is produced by a single laser beam of frequency ω and linear polarization i:

$$I_j^{(2\omega)} \propto \left(I_i^{(\omega)} \right)^2,$$

which means that the second harmonic intensity is proportional to the square of the fundamental intensity.

(3) Apparently, from Eq. (11.7), one would be able to increase the second harmonic intensity by increasing L because of the L^2-dependence. But one should beware that the function

$$\frac{\sin^2 \left(\frac{1}{2}(\Delta k)L \right)}{\left(\frac{1}{2}(\Delta k)L \right)^2} \tag{11.10}$$

overwhelms the L^2-dependence as can be seen in Fig. 11.1 in which the function $\text{sinc}^2 x$ is shown. Mathematically:

$$\text{sinc}^2 x \equiv \frac{\sin^2 \left(\frac{\pi x}{b} \right)}{(\pi x / b)^2} \tag{11.11}$$

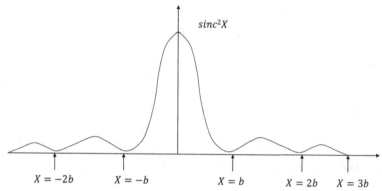

Fig. 11.1. Schematic representation of a sinc²x function. The drawing is not to scale.

so that expression (11.10) becomes Eq. (11.11) if
or

$$\pi x/b = \frac{1}{2}(\Delta k)L,$$

$$x = \frac{\Delta k}{2\pi}Lb. \tag{11.12}$$

As shown in Fig. 11.1, $\text{sinc}^2 x$ becomes zero periodically at:

$$x = \pm mb, m = 1, 2, 3, \ldots$$

Thus, the intensity of the second harmonic oscillates periodically. However, only the central peak (see Fig. 11.1) is significant, and we define the length L at which the first zero occurs as the *coherence length* l_{c},

$$\text{i.e.,} \quad L = l_{\text{c}} \quad \text{at} \quad x = b,$$

$$\text{i.e.,} \quad \frac{\Delta k}{2\pi}l_{\text{c}}b = b,$$

$$l_{\text{c}} = \frac{2\pi}{\Delta k}, \tag{11.13}$$

i.e., the second harmonic intensity is significant only within the coherence length l_{c} and is maximum when

$$x = 0, \quad (\text{see Fig. 11.1})$$

or when

$$\Delta k = 0 \tag{11.14}$$

for a finite L (see Eq. (11.12)). Using Eq. (11.8):

$$\Delta k \equiv k_j^{(2\omega)} - k_i^{(\omega)} - k_l^{(\omega)} = 0. \tag{11.15}$$

This is the very important *phase matching* condition.

11.3   Phase Matching

The phase matching condition (Eq. (11.15)) is usually the most critical one for ensuring an efficient second harmonic conversion. This is true not only in the low conversion efficiency approximation but also in the general case in which the fundamental laser beam is significantly depleted (turning into a second harmonic). We examine this in more detail. Since in general:

$$k = \frac{\omega}{c} n^{(\omega)},$$

Eq. (11.15) becomes:

$$\frac{2\omega}{c} n_j^{(2\omega)} - \frac{\omega}{c} n_i^{(\omega)} - \frac{\omega}{c} n_l^{(\omega)} = 0,$$

where $n_k^{(w)}$ is the index of refraction of a plane wave of frequency ω and polarized in the k-direction, with $k = i, j$ or l,

i.e., $$\left[n_j^{(2\omega)} - n_i^{(\omega)} \right] + \left[n_j^{(2\omega)} - n_l^{(\omega)} \right] = 0. \tag{11.16}$$

This yields two solutions. The first is when $i = l$, i.e., the fundamental frequency comes from the same pump laser generating a second harmonic wave of polarization $j (j = 1, 2)$. Thus, Eq. (11.16) yields:

$$n_j^{(2\omega)} - n_i^{(\omega)} = 0, \quad (\text{Type I phase matching}). \tag{11.17}$$

The second is when $i \neq l$, i.e., one employs two laser beams at the same fundamental frequency but with orthogonal linear polarizations. Note: $i, l = 1, 2$ so that when $i \neq l$, one of them must be equal to 1-polarized in the x-direction, and the other equal to 2-polarized in the y-direction. Equation (11.16) thus yields

$$n_j^{(2\omega)} = \frac{1}{2}\left[n_1^{(\omega)} + n_2^{(\omega)}\right], \quad \text{(Type II phase matching)}, \quad (11.18)$$

where we have explicitly expressed i and l as 1 and 2.

Case 1: Type I phase matching
Using one linearly polarized laser beam to generate second harmonic radiation, Eq. (11.17) is valid:

$$n_j^{(2\omega)} - n_i^{(\omega)} = 0. \qquad (11.19)$$

(a) If $j = i$, it means that the polarization of the second harmonic is parallel to that of the fundamental pump beam. If the medium is isotropic, because of the normal dispersion of the medium,

$$n_j^{(2\omega)} \neq n_i^{(\omega)}.$$

For example, most types of glass exhibit

$$n_j^{(2\omega)} > n_i^{(\omega)} \qquad (11.20)$$

in the near IR to the near UV region. Thus, it is almost impossible to satisfy Eq. (11.19) in an isotropic medium.

If the medium is anisotropic, $j = i$ means that the second harmonic and the fundamental are either both ordinary or extraordinary waves in the medium (using uniaxial crystal as an example). If they are both ordinary waves, their index ellipsoids are concentric spheres of radii $n^{(\omega)}$ and $n^{(2\omega)}$ and they will never be equal in general because of the dispersion (Eq. (11.20)). Thus, Eq. (11.19) cannot be satisfied. If the two waves are both extraordinary waves, their index ellipsoids are two concentric ellipsoids without any intersection between them. Again, $n^{(2\omega)}$ cannot be made equal to $n^{(\omega)}$ due to dispersion (Eq. (11.20)).

In conclusion, if $j = i$, the index matching condition (11.19) cannot be satisfied in both isotropic and anisotropic media.

(b) If $j \neq i$, Eq. (11.19) still cannot be satisfied in an isotropic medium because

$$n_j^{(\omega)} \equiv n_i^{(\omega)} \neq n_j^{(2\omega)}. \qquad (11.21)$$

We are left with the possibility of using an anisotropic medium. Using a uniaxial medium as an example, one can choose a direction in the

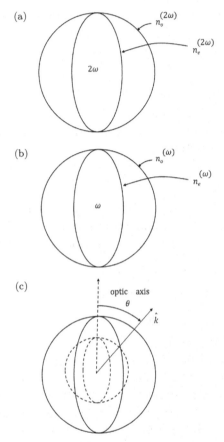

Fig. 11.2.   Phase matching in a negative uniaxial crystal.

medium such that the second harmonic propagates as an ordinary wave and the fundamental as an extraordinary wave (or vice versa) with equal velocity (hence equal indices, satisfying Eq. (11.19)). Figures 11.2 and 11.3 illustrate the above statement.

Figure 11.2 shows the surfaces of wave normals of a negative uniaxial medium. (a) corresponds to the surfaces of the wave normals at frequency 2ω, and (b) at ω. In (c), the two are superimposed and we see that in the direction $\hat{k}$, the spherical surface of radius $n_o^{(\omega)}$ of the ordinary wave of o frequency ω intersects the ellipsoid of index $n_e^{(2\omega)}(\theta)$ (θ is the angle between the optic axis and $\hat{k}$) of the extraordinary wave of frequency 2ω. At the intersection point, $n_o^{(\omega)} = n_e^{(2\omega)}(\theta)$,

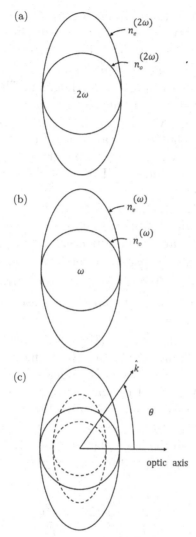

Fig. 11.3.   Phase matching in a positive uniaxial crystal.

thus satisfying Eq. (11.19). Physically speaking, it means that the second harmonic wave generated inside the anisotropic medium will have an ordinary and an extraordinary component both propagating in the $\hat{k}$-direction. The ordinary component will "see" an index $n_o^{(2\omega)}$ which will never be equal to $n_o^{(\omega)}$ or $n_e^{(\omega)}$ of the fundamental wave of frequency ω. Hence, the ordinary component of the second harmonic wave will quickly die out (or not be amplified) because of the phase mismatch (i.e., not satisfying Eq. (11.19)). The extraordinary component "sees" an index $n_e^{(2\omega)}(\theta)$ equal to $n_o^{(\omega)}$ of the fundamental. Note that their polarizations are orthogonal. The phase matching condition is thus satisfied and the second harmonic is amplified as it propagates at the same velocity as the pump beam (fundamental). Experimentally, one should thus align the fundamental laser beam in the direction of $\hat{k}$, i.e., the anisotropic crystal should be properly cut to permit an easy alignment. The angle θ can be calculated. We shall come back to this.

Similarly, Fig. 11.3 shows the surfaces of wave normals of a positive uniaxial medium. Again, (a) represents the surfaces of wave normals at 2ω and (b) at ω. The superposition of the two shows that in the direction $\hat{k}$, the propagation velocity (and hence indices) of the ordinary wave at 2ω (index $n_o^{(2\omega)}$) and the extraordinary wave at ω (index $n_e^{(\omega)}(\theta)$) are equal, satisfying the phase matching condition (Eq. (11.19)).

$n_e^{(\omega)}(\theta)$ (Fig. 11.3) or $n_e^{(2\omega)}(\theta)$ (Fig. 11.2) can be calculated easily in the case of a uniaxial crystal. Referring back to Section 6.3 and Fig. 6.12, a plane EM wave of frequency ω propagating in the direction $\hat{k}$ inside a uniaxial crystal generates two waves, ordinary and extraordinary, traveling in the $\hat{k}$-direction, with electric vectors $\boldsymbol{E_o}$ and $\boldsymbol{E_e}$ ($\boldsymbol{E_o} \perp \boldsymbol{E_e}$) and indices n_o and $n_e(\theta)$, respectively. Note that $\hat{k}$ is constrained in the $y-z$ plane without loss of generality and $n_e(\theta)$ changes from n_o at $\theta = 0$ to n_e at $\theta = \frac{\pi}{2}$. From Fig. 6.12, one sees that

$$\frac{1}{n_e^2(\theta)} = \frac{\cos^2\theta}{n_0^2} + \frac{\sin^2\theta}{n_e^2}, \qquad (11.22)$$

the derivation is left as an exercise to the reader.

In the case of a negative uniaxial crystal, one needs to calculate $n_e^{(2\omega)}(\theta)$. Equation (11.22) becomes:

$$\frac{1}{\left[n_e^{(2\omega)}(\theta)\right]^2} = \frac{\cos^2\theta}{\left[n_o^{(2\omega)}\right]^2} + \frac{\sin^2\theta}{\left[n_e^{(2\omega)}\right]^2}. \qquad (11.23)$$

At $\theta = \theta_{pm}$, the phase matching condition is assumed to be valid, and Eq. (11.19) becomes:

$$n_o^{(\omega)} = n_e^{(2\omega)}(\theta_{pm}) \tag{11.24}$$

θ_{pm} is the phase matching angle. Substituting Eq. (11.24) into (11.23), we obtain

$$\frac{1}{\left[n_0^{(\omega)}\right]^2} = \frac{\cos^2\theta_{pm}}{\left[n_o^{(2\omega)}\right]^2} + \frac{\sin^2\theta_{pm}}{\left[n_e^{(2\omega)}\right]^2} \tag{11.25}$$

with $n_o^{(\omega)}$, $n_o^{(2\omega)}$ and $n_e^{(2\omega)}$ known. Equation (11.25) can be solved for the angle θ_{pm} (exercise):

$$\sin^2\theta_{pm} = \frac{\left[n_0^{(\omega)}\right]^{-2} - \left[n_0^{(2\omega)}\right]^{-2}}{\left[n_e^{(2\omega)}\right]^{-2} - \left[n_o^{(2\omega)}\right]^{-2}}, \quad \text{(negative uniaxial crystal)}.$$

$$\tag{11.26}$$

Similarly, in the case of a positive uniaxial crystal, we need to calculate $n_e^{(\omega)}(\theta)$. Equation. (11.22) becomes:

$$\frac{1}{\left[n_e^{(\omega)}(\theta)\right]^2} = \frac{\cos^2\theta}{\left[n_o^{(\omega)}\right]^2} + \frac{\sin^2\theta}{\left[n_e^{(\omega)}\right]^2}. \tag{11.27}$$

Using the phase matching condition at $\theta = \theta_{pm}$,

$$n_0^{(2\omega)} = n_e^{(\omega)}(\theta_{pm}), \tag{11.28}$$

one obtains (exercise):

$$\sin^2\theta_{pm} = \frac{\left[n_o^{(2\omega)}\right]^{-2} - \left[n_o^{(\omega)}\right]^{-2}}{\left[n_e^{(\omega)}\right]^{-2} - \left[n_o^{(\omega)}\right]^{-2}}, \text{(positive uniaxial crystal)}. \tag{11.29}$$

Case 2 Type II phase matching

Using two parallel pump beams with orthogonal linear polarizations to generate second harmonic radiation, Eq. (11.18) is valid:

$$n_j^{(2\omega)} = \frac{1}{2}\left[n_1^{(\omega)} + n_2^{(\omega)}\right]. \tag{11.30}$$

The situation becomes much more complicated. We shall consider simple situations only. Again, using a uniaxial crystal as an example, the polarization directions 1 and 2 in Eq. (11.30) are assumed to be those of the ordinary and extraordinary waves for the sake of simplicity. Thus, Eq. (11.30) becomes:

$$n_j^{(2\omega)} = \frac{1}{2}\left[n_o^{(\omega)} + n_e^{(\omega)}\right]. \tag{11.31}$$

The propagation direction $\hat{k}$ of the three waves (two pump beams and the second harmonic) is assumed collinear, and $\hat{k}$ makes an angle θ, in general, with the optic axis (see Figs. 11.2, 11.3 and 6.12). There could now be several ways of fulfilling Eq. (11.31) depending on the indices of the anisotropic material. First of all, one has, in general,

$$n_j^{(2\omega)}(\theta_{\mathrm{pm}}) = \frac{1}{2}\left[n_o^{(\omega)} + n_e^{(\omega)}(\theta_{\mathrm{pm}})\right], \tag{11.32}$$

where θ_{pm} is the phase matching angle. It is easy to imagine that some material could satisfy Eq. (11.32) at $\theta_{\mathrm{pm}} = 0$. Under such a condition, all the waves become ordinary and

$$n_j^{(2\omega)}(\theta_{\mathrm{pm}} = 0) = n_o^{(2\omega)}$$

$$n_e^{(\omega)}(\theta_{\mathrm{pm}} = 0) = n_o^{(\omega)}$$

and Eq. (11.32) becomes:

$$n_o^{(2\omega)} = \frac{1}{2}\left[n_o^{(\omega)} + n_e^{(\omega)}\right] = n_o^{(\omega)}, \tag{11.33}$$

regardless of polarization directions. But this condition is impossible due to dispersion (cf. Eq. (11.20)). Thus, we are left with two more possibilities for $\theta_{\mathrm{pm}} \neq 0$, namely,

$$n_j^{(2\omega)}(\theta_{\mathrm{pm}}) = \begin{cases} n_o^{(2\omega)}(\theta_{\mathrm{pm}}) \\ n_e^{(2\omega)}(\theta_{\mathrm{pm}}) \end{cases}, \tag{11.34}$$

i.e., the second harmonic could be an ordinary $n_o^{(2\omega)}(\theta_{\mathrm{pm}})$ or an extraordinary $n_e^{(2\omega)}(\theta_{\mathrm{pm}})$ wave.

If the second harmonic is an ordinary wave, we have, from Eq. (11.32):

$$n_{\mathrm{o}}^{(2\omega)}(\theta_{\mathrm{pm}}) = \frac{1}{2}\left[n_{\mathrm{o}}^{(\omega)} + n_{\mathrm{e}}^{(\omega)}(\theta_{\mathrm{pm}})\right]. \tag{11.35}$$

But $n_{\mathrm{o}}^{(2\omega)}(\theta_{\mathrm{pm}})$ is constant, independent of θ_{pm} because, by definition, the surface of the wave normal of the ordinary wave is a sphere and thus, the wave propagates at the same velocity in all directions, i.e., constant index in all directions. Equation (11.35) is thus not satisfied because $n_{\mathrm{e}}^{(\omega)}(\theta_{\mathrm{pm}})$ varies with θ_{pm}. We are left with the last possibility; namely, the second harmonic must be an extraordinary wave in order to satisfy the phase matching condition. Equation (11.32) becomes:

$$n_{\mathrm{e}}^{(2\omega)}(\theta_{\mathrm{pm}}) = \frac{1}{2}\left[n_{\mathrm{o}}^{(\omega)} + n_{\mathrm{e}}^{(\omega)}(\theta_{\mathrm{pm}})\right]. \tag{11.36}$$

To solve for θ_{pm}, we use Eq. (11.22) for both ω and 2ω:

$$\frac{1}{\left[n_{\mathrm{e}}^{(2\omega)}(\theta_{\mathrm{pm}})\right]^2} = \frac{\cos^2\theta_{\mathrm{pm}}}{\left[n_{\mathrm{o}}^{(2\omega)}\right]^2} + \frac{\sin^2\theta_{\mathrm{pm}}}{\left[n_{\mathrm{e}}^{(2\omega)}\right]^2}, \tag{11.37}$$

$$\frac{1}{\left[n_{\mathrm{e}}^{(\omega)}(\theta_{\mathrm{pm}})\right]^2} = \frac{\cos^2\theta_{\mathrm{jm}}}{\left[n_{\mathrm{o}}^{(\omega)}\right]^2} + \frac{\sin^2\theta_{\mathrm{pm}}}{\left[n_{\mathrm{e}}^{(\omega)}\right]^2}. \tag{11.38}$$

By substituting Eqs. (11.37) and (11.38) into (11.36), one obtains an equation with θ_{pm} as the only unknown variable. We shall not go into the details here.

Summary (1): Type I phase matching condition

Both the linearly polarized pump and second harmonic waves propagate in the same direction $\hat{k}$ that makes an angle θ_{pm} (the phase matching angle) with the optic axis. For a negative uniaxial medium:

$$n_{\mathrm{o}}^{(\omega)} = n_{\mathrm{e}}^{(2\omega)}(\theta_{\mathrm{pm}}), \quad (\text{Eq. (11.24)})$$

i.e., the pump beam is an ordinary wave and the second harmonic is an extraordinary wave both propagating in the direction $\hat{k}$ at the same velocity (in phase). The experimental schematic is given in

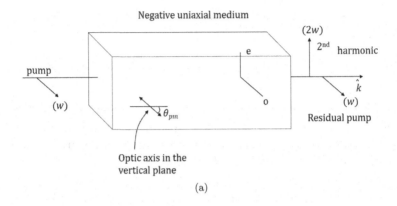

Negative uniaxial medium

(a)

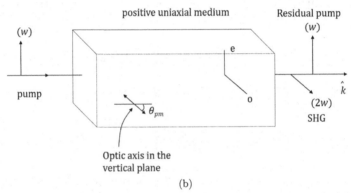

positive uniaxial medium

(b)

Fig. 11.4. Type I phase matching in the SHG in (a) negative, and (b) positive uniaxial media.

Fig. 11.4(a). The medium is cut such that $\theta_{\rm pm}$ (see Eq. (11.26)) can be easily aligned. For a positive uniaxial medium:

$$n_{\rm o}^{(2\omega)} = n_{\rm e}^{(\omega)}(\theta_{\rm pm}), \quad \text{(Eq. (11.28))}$$

i.e., the pump beam is an extraordinary wave and the second harmonic, an ordinary wave, both propagating in the direction $\hat{k}$ at the same velocity (in phase). The angle $\theta_{\rm pm}$ is given by Eq. (11.29). The experimental schematic is shown in Fig. 11.4(b).

Summary (2): Type II phase matching condition

Two pump beams with orthogonal linear polarizations are used — one is an ordinary wave, the other, an extraordinary wave. The generated second harmonic is an extraordinary wave. All the waves propagate in the same direction $\hat{k}$, making an angle $\theta_{\rm pm}$ with respect

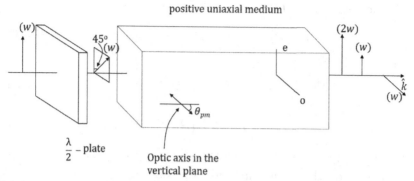

Fig. 11.5.   One practical way for type II phase matching in the SHC in a uniaxial crystal.

to the optic axis. Equation (11.36) has to be satisfied. This means that the "mean" velocity of the combined pump waves is equal to the velocity of the second harmonic wave. Figure 11.5 shows one practical way of such phase matching starting from one pump beam. The pump beam first passes through a half-wave plate such that the incident linearly (vertically) polarized pimp beam becomes polarized at 45° with respect to the vertical axis (e-axis in the crystal) (see Chapter 7). In entering the second harmonic medium, the incident wave is decomposed into ordinary and extraordinary waves. These two waves thus fulfill the two pump beams' condition of orthogonal polarizations. At the output, there are three waves, the two pump waves (now phase shifted with each other) and the second harmonic wave polarized vertically. The two pump waves combine, forming a resultant wave of elliptic polarization in general.

The output waves of frequencies 2ω and ω in Figs. 11.4 and 11.5 can be separated by various means. For instance, one can use a prism to separate them or a thin film dichroic beam splitter that transmits preferentially one frequency and reflects the other.

11.4   Sum- and Difference-Frequency Generation

When the pump field consists of two different frequency components ω_1 and ω_2, i.e., $\omega_1 \neq \omega_2$, and passes through a nonlinear uniaxial crystal, new radiation waves at the sum-frequency of $\omega_3 = \omega_1 + \omega_2$ and the difference-frequency of $\omega_3 = \omega_1 - \omega_2$ can be respectively generated in certain conditions.

According to Eq. (11.4), we have:

$$E(z,t) = E_1^{(\omega_1)}(z,t) + E_2^{(\omega_2)}(z,t) \tag{11.39}$$

where

$$E_1^{(\omega_1)}(z,t) = \frac{1}{2}E_1^{(\omega_1)}(z)\{e^{i(\omega_1 t - kz)} + c.c.\}, \tag{11.40}$$

$$E_2^{(\omega_2)}(zt) = \frac{1}{2}E_2^{(\omega_2)}(z)\{e^{i(\omega_2 t - kz)} + c.c.\}. \tag{11.41}$$

Substituting Eqs. (11.39) to (11.41) into (11.3) and keeping only the second term on the right hand side, one obtains:

$$P_i^{(\omega_1,\omega_2)}(z,t) \propto E(z,t)E^*(z,t), \tag{11.42}$$

$$\propto E_1^2 \exp(2i\omega_1 t) + E_1^{*2}\exp(-2i\omega_1 t), \tag{11.43}$$

$$+ E_2^2 \exp(2i\omega_2 t) + E_2^{*2}\exp(-2i\omega_2 t), \tag{11.44}$$

$$+ E_1 E_2 \exp(i(\omega_1 + \omega_2)t)$$

$$+ E_1^* E_2^* \exp(-i(\omega_1 + \omega_2)t), \tag{11.45}$$

$$+ E_1 E_2 \exp(i(\omega_1 + \omega_2)t)$$

$$+ E_1^* E_2^* \exp(-i(\omega_1 + \omega_2)t), \tag{11.46}$$

where the terms (11.43) and (11.44) represent the second harmonic generations of the pump field at the two frequencies of ω_1 and ω_2, respectively, and the terms (11.45) and (11.46) result in the new radiation waves at the sum-frequency of $\omega_3 = \omega_1 + \omega_2$ and the difference-frequency of $\omega_3 = \omega_1 - \omega_2$, respectively.

Moreover, Eqs. (11.40) and (11.41) can be rewritten as:

$$E_1^{(\omega_1)}(z,t) \propto E_1^{(\omega_1)}(z)e^{-ik_1 z}, \tag{11.47}$$

$$E_2^{(\omega_2)}(zt) \propto E_2^{(\omega_2)}(z)e^{-ik_2 z}. \tag{11.48}$$

Therefore, according to Eq. (11.42), one obtains the expression for the sum-frequency term:

$$P_i^{(\omega_1,\omega_2)}(z,t) \propto E_1^{(\omega_1)}(z)E_2^{(\omega_2)}e^{-i(k_1+k_2)z}. \tag{11.49}$$

Comparing Eq. (11.49) with the corresponding sum-frequency field expression:

$$E_3^{(\omega_3)}(zt) \propto E_3^{(\omega_3)}(z)e^{-ik_3 z}, \tag{11.50}$$

one can find that only in the condition of

$$\Delta k \equiv \boldsymbol{k}_3 - (\boldsymbol{k}_1 + \boldsymbol{k}_2) = 0, \tag{11.51}$$

the phase matching is achieved. When $\omega_1 = \omega_2$ and $k_1 = k_2$, the result is exactly the same as that of Eq. (11.15). That is, the second harmonic generation can be regarded as a degenerate sum-frequency generation, which means that these two processes are the same type of second-order nonlinear processes.

Substituting $k = \frac{\omega}{c} n^{(\omega)}$ into Eq. (11.51), one obtains

$$\frac{\omega_3}{c} n_j^{(\omega_3)} - \frac{\omega_2}{c} n_i^{(\omega_2)} - \frac{\omega_1}{c} n_l^{(\omega_1)} = 0,$$

i.e.,

$$\omega_3 n_j^{(\omega_3)} - \omega_2 n_i^{(\omega_2)} - \omega_1 n_l^{(\omega_1)} = 0. \tag{11.52}$$

Clearly, the nonlinear processes of second harmonic and sum-frequency generation satisfy the energy and momentum conservations.

$$h\omega_3 = h\omega_1 + h\omega_2, \tag{11.53}$$

$$\boldsymbol{k}_3 = \boldsymbol{k}_1 + \boldsymbol{k}_2 \tag{11.54}$$

Similarly, one can obtain the phase matching expressions for the difference-frequency generation, i.e.,

$$\omega_3 = \omega_1 - \omega_2, \tag{11.55}$$

$$\Delta k \equiv \boldsymbol{k}_3 - (\boldsymbol{k}_1 - \boldsymbol{k}_2) = 0, \tag{11.56}$$

in which the conservation of energy and momentum hold. Since the phase matching of sum and difference-frequency generation in a uniaxial crystal can be obtained in a similar manner as that in the second harmonic generation, we shall not go into the details here (exercise).

To intuitively see the second-order nonlinear processes in a quantum point of view, in Fig. 11.6, the quantum transition diagrams for the second-harmonic, sum- and difference-frequency generations are presented. For second-harmonic generation, an atom/ion or a molecule in a second-order nonlinear crystal is excited to an intermediate state (1) with the annihilation of a photon with the energy $h\omega$, which is then excited to the intermediate state (3) with the annihilation of another photon with the same energy of $h\omega$. When the

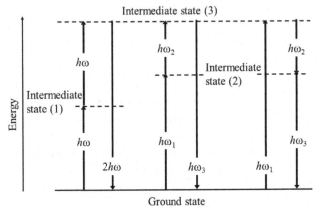

Fig. 11.6. Transition diagrams for the second-harmonic, sum- and difference-frequency generation in a nonlinear uniaxial crystal.

excited particle goes back to the ground state without any delay, a photon with the energy of $2\hbar\omega$ is emitted. A similar process takes place for the sum-frequency generation, which satisfies the condition of $\hbar\omega_3 = \hbar\omega_1 + \hbar\omega_2$, in which only the energies of the two annihilated photons are different. But in the difference-frequency generation, the particle in the nonlinear crystal is excited directly to the intermediate state (3) with the annihilation of a photon of $\hbar\omega_1$. It then decays to the intermediate state (2) with the emission of a photon of $\hbar\omega_2$ and then returns without any delay to the ground state with the emission of another photon of $\hbar\omega_3$ where $\hbar\omega_3 = \hbar\omega_1 - \hbar\omega_2$. This means that the optical energy transfers from the high-frequency photon to the other two low-frequency photons; that is, the annihilation of one high-frequency photon creates two low-frequency photons. In this case, if one puts the crystal in a cavity with the incident light containing the two waves of ω_1 and ω_2, the wave ω_2 will be amplified (called signal light); meanwhile, a new wave of ω_3 (called idler light) will be simultaneously generated. This is a typical optical parametric amplification process.

11.5 Third and High Harmonic Generation

We now briefly discuss the third-harmonic generation in an anisotropic medium based on the third-order nonlinear polarization process, i.e., we only consider the third term on the right hand of Eq. (11.3). With the equation of the pumping wave in Eqs. (11.4) and (11.3), the expression for the intensity $I_j^{(3\omega)}$ of the third harmonic radiation under an additional assumption of low conversion efficiency, i.e., $\frac{dE_i^{(\omega)}}{dz} \simeq 0$, can be obtained:

$$I^{(3\omega)} = \frac{9\omega^2}{\epsilon_0^2 c^4 n_\omega^3 n_\omega}|x_{ijkl}|^2 (I^{(\omega)})^2 L^2 \frac{\sin^2(\frac{1}{2}\Delta kL)}{(\frac{1}{2}\Delta kL)^2}. \qquad (11.57)$$

Similar to Eq. (11.13), the coherence length l_c for the third-harmonic generation can be written as:

$$l_c = \frac{2\pi}{\Delta k} \qquad (11.58)$$

and the phase matching condition is:

$$\Delta k = k_j^{(3\omega)} - k_i^{(\omega)} - k_l^{(\omega)} - k_k^{(\omega)} = 0. \qquad (11.59)$$

Similar to the deduction of the phase matching in second-harmonic generation, one can obtain the conditions for the phase matching of different light polarization states in, for example, a negative uniaxial crystal.

$$n_e^{(3\omega)}(\theta_{pm}) = n_o^{(\omega)} \quad \text{(Type I)}, \qquad (11.60)$$

$$n_e^{(3\omega)}(\theta_{pm}) = \frac{1}{3}\left[n_o^{(\omega)} + n_0^{(\omega)} + n_e^{(\omega)}(\theta_{pm})\right], \quad \text{(Type II)},$$

$$(11.61)$$

where θ_{pm} is the phase matching angle. We shall not go into the details to deduce θ_{pm} here.

Since the third-order susceptibility is much smaller than the second-order susceptibility, one uses two second-order nonlinear crystals — one is used to produce the second harmonic wave of 2ω, and the other is used to generate the sum-frequency (3ω) of the fundamental pump light wave (ω) and the second harmonic light (2ω) — to generate a third harmonic wave, instead of directly producing the third harmonic radiation wave (3ω) of a pump light (ω) through the third-order nonlinear process in a nonlinear medium.

Moreover, when the pump field consists of three different frequency components ω_1, ω_2 and ω_3, the electric field of the pump light can be expressed as:

$$E(z,t) = E_1^{(\omega_1)}(z,t) + E_2^{(\omega_2)}(z,t) + E_2^{(\omega_3)}(z,t) \tag{11.62}$$

Substituting into Eq. (11.3), one can obtain different frequency waves from the third term on the right hand of Eq. (11.3), such as:

$$(\omega_i + \omega_i + \omega_i), (i = 1, 2, 3), \tag{11.63}$$

$$(\omega_1 + \omega_2 + \omega_3), \tag{11.64}$$

$$(\omega_1 + \omega_2 - \omega_3), \tag{11.65}$$

where frequencies of these three waves can be the same or different, depending on the phase matching condition. These third-order nonlinear processes give rise to many interesting phenomena, such as four-wave mixing, resonance enhancement of third-harmonic generation, Raman-enhanced four-wave mixing such as coherent Stokes- and anti-Stokes scattering, phase conjugation, frequency-degenerate four-wave mixing, and the optical Kerr effect, which provides a toolbox for various laser spectroscopy techniques and laser applications.

It should be also pointed out that when the pump light, such as high-peak-power femtosecond laser pulses, is very strong and passes the nonlinear medium, the higher-order nonlinear terms in Eq. (11.3) cannot be negligible. In that case, higher-order harmonic generation in an extremely short wavelength, even at a hard X-ray region, can be achieved. In addition, with the generation of different orders of high harmonics in the spectrum, an ultrashort pulse in an attosecond $(10^{-18}\,\text{s})$ time scale enables to be obtained. We shall not go into the details here, and the interested reader can consult the available texts in the market.

Chapter 12

Mode Locking and Carrier-Envelope Phase Locking

We arbitrarily define a *short laser pulse* as one whose duration (*full width at half maximum (FWHM)*) is of the order of a few tens of nanoseconds and shorter. The lower limit depends on the wavelength of the laser. The shortest laser pulse to date is 43 as (attoseconds or 10^{-18} seconds) in duration in the ultraviolet region. In terms of an optical cycle, it lasts only a few optical cycles. The reason why an attosecond pulse is in the UV region is that the pulse duration must contain more than one optical cycle. The duration of one optical cycle for 300-nm wave corresponds to ~ 1 fs (1000 as). For pulse duration in the scope of attosecond (below 1 fs), the corresponding wavelength must be shorter than 300 nm, thus, in the UV region. We have already qualitatively described Q-switching and mode locking in Chapters 8 and 9. In this chapter, mode locking and carrier-envelope phase (CEP) locking will be introduced. Mode locking can shorten the pulse duration to the scope of tens of attoseconds with the help of high harmonic generation, which is the most representative method today to obtain ultrashort pulses from lasers. Meanwhile, CEP locking has become an important tool in precision optical frequency measurements. This chapter is structured with four sections except the Closing mark. They are as follows, *Principle of mode locking, Mode locking techniques, Carrier-envelope phase, and Carrier-envelope phase measurement and stabilization.*

12.1 Principle of Mode Locking

With the advent of lasers, the output pulse durations have continuously decreased from the microsecond domain with the free running of a solid laser to a nanosecond regime with Q-switching, and finally to a picosecond or femtosecond regimes with mode locking. Mode-locked lasers generate a train of repetitive ultrashort optical pulses by fixing the relative phases of all of the lasing longitudinal modes (see Section 5.9 for the definition of longitudinal modes).

If we assume that the generated longitudinal mode corresponds to a certain transverse mode and the beam propagates along the z-axis, the time-dependent electric field of the q-th longitudinal mode at position $z = 0$ can be expressed as:

$$E_q(t) = E_q e^{i(\omega_q t + \varphi_q)}, \tag{12.1}$$

where the amplitude, angular frequency, and initial phase are indicated by E_q, ω_q and φ_q, respectively. The total electric field, by superposing all longitudinal modes is:

$$E(t) = \sum_q E_q e^{i(\omega_q t + \varphi_q)}. \tag{12.2}$$

Since there is no coupling for frequencies and initial phases among all the longitudinal modes, the superposition of all longitudinal modes in Eq. (12.2) is incoherent, which means that the output intensity fluctuates.

Mode locking forces all the longitudinal modes in a resonator to oscillate synchronously in the frequency domain with a constant locked interval between adjacent modes and a common initial phase. Thereby, it results in a repetitive train of ultrashort optical pulses in the time domain. Assuming that there are $2N + 1$ longitudinal modes in a resonator, which is $q = -N, -(N-1), \ldots, 0, \ldots, (N-1), N$, the angular frequencies of the q-th longitudinal modes is given by:

$$\omega_q = \omega_0 + q\Omega, \tag{12.3}$$

where ω_0 is the central frequency, and the difference between adjacent angular frequencies is $\Omega = (2\pi)(c/2L)$. Here, L is the optical length of the laser cavity and $c/2L$ is the inverse of the round trip time of light propagation in the cavity or the frequency difference between

two adjacent longitudinal modes. Therefore, it can be obtained from
Eq. (12.2):

$$E(t) = \sum_{-N}^{N} E_q \exp[i(\omega_0 + q\Omega)t + \varphi_q]. \tag{12.4}$$

Assume that all longitudinal modes have equal amplitude, meaning
$E_q = E_0$. Assume also that the adjacent modes have a fixed interval
in the frequency domain, and every mode oscillates with a common
initial phase ($\varphi_q = 0$). By trigonometric series, Eq. (12.4) can be
derived as:

$$E(t) = E_0 \left(\sum_{-N}^{N} e^{iq\Omega t} \right) e^{i\omega_0 t} = \left\{ E_0 \frac{\sin \left[\frac{1}{2}(2N+1)\Omega t \right]}{\sin \frac{1}{2}(\Omega t)} \right\} e^{i\omega_0 t}.$$

$$\tag{12.5}$$

The above formula shows the amplitude by synthesizing $(2N+1)$ longitudinal modes in a mode-locked resonator. In this case, the accumulated intensity is:

$$I(t) \propto E_0^2 \frac{\sin^2 \left[(2N+1)\frac{\Omega t}{2} \right]}{\sin^2 \left(\frac{\Omega t}{2} \right)}. \tag{12.6}$$

Figure 12.1 is plotted based on Eq. (12.6). From Fig. 12.1, we can
see that mode locking technology makes laser energy highly concentrated in time by locking the relative phases of all longitudinal modes.
In Fig. 12.1, when $\Omega t/2 = m\pi$, $m = 0, 1, 2 \ldots$, the accumulated light
intensity $I(t)$ reaches its maximal value.

$$I(t) \propto \lim_{\frac{\Omega t}{2} \to m\pi} \frac{E_0^2 \sin^2 \left[(2N+1)\frac{\Omega t}{2} \right]}{\sin^2 \left(\frac{\Omega t}{2} \right)} = (2N+1)^2 E_0^2. \tag{12.7}$$

From Eq. (12.7), with mode locking, the accumulated light intensity through interference (coherent superposition) of $(2N+1)$ modes
is directly proportional to $(2N+1)^2$. Without mode locking, the
light intensity is the incoherent superposition of $(2N+1)$ modes.
Therefore, we can conclude that in the case of mode locking, higher
intensity can be achieved with more longitudinal modes. This can be
done by either lengthening the cavity or widening the fluorescence

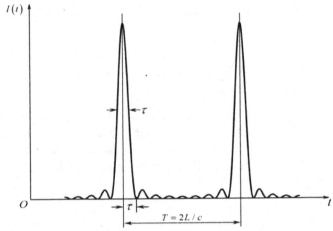

Fig. 12.1.   A repetitive train of ultrashort optical pulses with locked longitudinal modes.

linewidth (see below for justification). The time interval between adjacent pulses can be derived from Eq. (12.5).

$$T = 2\pi/\Omega = 2L/c, \tag{12.8}$$

which is equal to the round trip time of an oscillating pulse. It means that an output pulse is only produced once per cavity round trip.

The pulse duration (τ) is defined as the time interval between the highest pulse peak and its nearest valley (see Fig. 12.1). In Eq. (12.6), t for the weakest light intensity $I(t)$ can be given when

$$\sin\left[(2N+1)\frac{\Omega t}{2}\right] = 0. \tag{12.9}$$

The weakest light intensity corresponds to the valleys in Fig. 12.1. The value of t in Eq. (12.9) can be derived as $t = \frac{2\pi}{(2N+1)\Omega}$. Therefore, the pulse duration can be expressed as:

$$\tau = \frac{2\pi}{(2N+1)\Omega} = \frac{T}{2N+1} = \frac{1}{\Delta v}, \tag{12.10}$$

where Δv is the bandwidth of a mode-locked laser. The above equation shows that the pulse duration is approximately the reciprocal of the width of the gain curve. Therefore, larger gain bandwidth is essential for generating ultrashort pulses. With the help of modern material sciences, gain media with large gain bandwidth have been developed, for instance, crystals of Nd:YAG, Ti:sapphire, CaF_2, etc.

12.2 Mode Locking Techniques

In mode locking, a laser resonator usually contains mode-locking devices (mode locker), either an active element (an optical modulator) or a passive element (a saturable absorber). Both of these can cause the formation of an ultrashort pulse circulating in the laser resonator. Mode-locking techniques include active mode locking, passive mode locking, synchronous mode locking, injection mode locking, and so on. The first two are most widely used for fabricating commercial mode-locked laser oscillator producing ultrashort pulses.

12.2.1 Active mode locking

For active mode locking, a modulator with the modulation frequency at $v = c/(2L)$ is inserted into the resonant cavity to manipulate the amplitude (cavity-loss modulation) or phase (phase modulation) of the oscillations in order to realize the synchronization of each longitudinal mode. Here, c is the speed of light in a vacuum, and L is the optical length of the cavity. The modulator can be an acousto-optic modulator, an electro-optic modulator, a semiconductor electro-absorption modulator, etc. If the modulation is synchronized with the resonator's round trip time, it can lead to the generation of ultrashort pulses.

As shown in Fig. 12.2(a), an electro-optic or acousto-optic cavity-loss modulator is employed by the resonator. The modulation frequency $v_m = c/2L$ (or for the modulation period $T_m = 2L/c$) is set to be equal to the interval between adjacent longitudinal modes. The modulator imposes strong loss outside the modulation period, thus, the optical path of the oscillator is normally blocked. However, once the modulation period comes, the modulator opens a "time gate" with much less loss lasting from several nanoseconds to hundreds of microseconds. At this time, the optical path of the oscillator is switched on. The principle of cavity-loss modulation can be explained from the perspective of mode coupling. The amplitude of

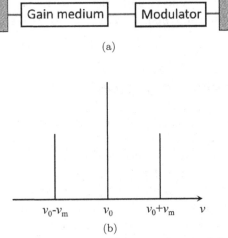

(a)

(b)

Fig. 12.2.   (a) Mode locking is achieved by cavity-loss modulation in a resonator. (b) The central and adjacent frequencies for the output of a mode-locked laser.

each longitudinal mode passing through the modulator changes periodically; it follows the periodical voltage applied to the modulator. If the central frequency at the gain curve is v_0, the mode at the central frequency first oscillates with the modulated electric field as:

$$
\begin{aligned}
\boldsymbol{E}(t) &= (\boldsymbol{E}_0 + \boldsymbol{E}_{\mathrm{m}} \cos 2\pi v_{\mathrm{m}} t) \cos 2\pi v_0 t \\
&= \boldsymbol{E}_0 (1 + M \cos 2\pi v_{\mathrm{m}} t) \cos 2\pi v_0 t,
\end{aligned}
\tag{12.11}
$$

where $M = E_m/E_o$ is the modulation of the amplitude, with its value depending on the amplitude of modulation. Equation (12.11) can be expanded as:

$$
\boldsymbol{E}(t) = \boldsymbol{E}_0 \cos 2\pi v_0 t + \boldsymbol{E}_0 \frac{M}{2} \cos 2\pi (v_0 - v_{\mathrm{m}}) t
$$

$$
+ \boldsymbol{E}_0 \frac{M}{2} \cos 2\pi (v_0 + v_{\mathrm{m}}) \boldsymbol{t}.
\tag{12.12}
$$

The above formula indicates that the oscillation is at the central frequency v_0 and two side bands ($v_0 \pm v_m$). The two side bands have the same initial phase as the central one. The modulation frequency v_m is equal to the frequency interval between the adjacent longitudinal modes. It means that in a resonator, once the central frequency v_0 is formed, two adjacent modes will start to circulate. For the same reason, the generated side bands $(v_0 \pm v_m)$ will produce another pair of side frequencies as well resulting in $(v_0 \pm 2v_m)$. In this sense, all longitudinal modes within the gain curve are excited eventually. Due to the initial phase locking of each longitudinal mode, the output optical energy is coherently superposed, meaning mode-locking.

Active mode locking can also be initiated through phase modulation. If an electro-optic phase modulator is employed by the resonator, the phase modulator can tune the electric field with a modulation frequency v_m. In this case, the resultant electric field is written as:

$$\mathbf{E}(t) = \mathbf{E}_0 \cos(2\pi v_0 t + \beta \sin 2\pi v_m t), \tag{12.13}$$

where β is the phase modulation depth. The above equation can be further expressed as functions of Bessel function, which is:

$$\begin{aligned}
E(t) = E_0[&J_0(\beta)\cos 2\pi v_0 t + J_1(\beta)\cos 2\pi(v_0 + v_m)t \\
&- J_1(\beta)\cos 2\pi(v_0 - v_m)t \\
&+ J_2(\beta)\cos 2\pi(v_0 + 2v_m)t + J_2(\beta)\cos 2\pi(v_0 - 2v_m)t \\
&+ J_3(\beta)\cos 2\pi(v_0 + 3v_m)t \\
&- J_3(\beta)\cos 2\pi(v_0 - 3v_m)t + J_4(\beta)\cos 2\pi(v_0 + 4v_m)t \\
&+ J_4(\beta)\cos 2\pi(v_0 - 4v_m)t + \cdots].
\end{aligned} \tag{12.14}$$

Equation (12.14) indicates that coherent phase modulation can also create the circulation of side bands $\nu_m(m = 0, 1, 2, \ldots)$ within the gain bandwidth, giving rise to mode locking.

12.2.2 Passive mode locking

In the case of passive mode locking, a thin cell containing a saturable absorber is employed. Figure 12.3 shows the relationship between the transmittance of a saturable absorber and the peak intensity of a pulse. The saturable absorber has smaller transmittance with lower intensity of the pulse and larger transmittance with stronger intensity.

When the pumping mechanism is activated, fluorescence from the gain medium sets in. The fluorescence noise travels back and forth in the resonator and grows in intensity. The intensity distribution

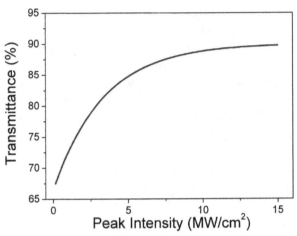

Fig. 12.3.   Transmittance of a saturable absorber versus the peak intensity of a pulse.

of the noise is random and contains many sharp peaks. When their intensity becomes strong enough, they will start to "bleach" totally or partially, the saturable absorber (see Fig. 12.3). Here, "bleach" means the saturation of the absorber. The higher intensity portions of the noise pass through the absorber with less loss. The lower intensity portions of the noise are absorbed and eventually disappear. Slowly, one peak emerges (not necessarily the one that was the most powerful to start with because of the relative timing with respect to bleaching) and is eventually the only one to lase.

Up to now, we have assumed that the gain is linear, which is almost the case with laser systems whose upper-level lifetimes are long (ruby, Nd). But for dye lasers, for instance, the active medium itself behaves nonlinearly and contributes as much as the saturable absorber to the ultra-short pulse generation process. It does so by sharpening the small pulse to a value much shorter than the recovery time of the absorber (τ_{rec}). The leading edge of the pulse saturates the gain in the active medium so that the tail is amplified less than the front, resulting in a shorter pulse. To summarize, we can distinguish two classes of passively mode-locked lasers: class I (e.g., ruby, neodymium), where the pulse width equals τ_{rec}, and class II (e.g., dyes) for which it is not limited by τ_{rec}.

12.2.3 Synchronous mode locking

In synchronous mode locking, the active medium of the laser is optically pumped by another mode-locked laser. The gain is modulated every $2L/c$ seconds. This is the round trip time of a pulse traveling in a cavity of optical length L. It is critical that the optical lengths of the resonators of the two lasers be matched; otherwise, the gain of the laser being pumped will not be modulated precisely every $2L/c$ seconds. This scheme offers certain unique advantages, like simplicity and two well-synchronized trains of short pulses at different wavelengths (the pump and the laser itself). It is worth noting that the optical pumping does not have to be coaxial with the optical axis of the laser being pumped. This scheme has been applied to dye lasers and fiber lasers. It is reliable and yields pulses almost as short as passive mode locking, provided that the pumping pulses are spaced precisely by $c/2L$ (or its multiple) in the frequency space. This is the most critical adjustment to perform.

12.2.4 Injection mode locking

This scheme requires a master laser and a slave laser. The master is mode-locked using one of the techniques described above. A single pulse is switched out of the train of pulses and injected into the slave having the same or similar spectral gain profile as the master laser has. The slave laser's cavity consists of an amplifier with similar gain bandwidth as that of the master laser. In this case, the injected pulse undergoes *regenerative amplification*. That is, the injected pulse, bouncing back and forth within the slave's cavity, is amplified each time it goes through the active medium and gives rise to an output pulse each time it strikes the output coupler of the slave. The master pulse has to be injected at the right time to dominate and grow more rapidly than the noise in the slave's cavity.

12.3 Carrier-envelope phase (CEP) Locking

Carrier-envelope phase (CEP) locking is one of the emphases of this chapter. CEP locking means stabilization of the pulse train in a mode-locked laser. CEP locking has demonstrated its applications in many fields. For instance, pulses with CEP locking are used as the seed pulse for pulses' coherent additions. Lasers with CEP locking are preferred by optical frequency metrology, which is connected with fundamental physical constants calibration, gravity wave detection, optical communication, optical ranging, etc. The purpose of this section is to look into the principles of CEP locking. It would be nice if our readers keep this question, "How could we lock the CEP of a mode-locked laser?" in mind when they go through Sections 12.3 and 12.4.

Before we start this section, let us go through the definitions of the related physical quantities. Meanwhile, the time-frequency correspondence and relation between the *CEP, repetition frequency*, and *offset frequency* are shown in Fig. 2.4, in order to explain the physical quantities clearly.

Group velocity: The velocity with which the envelope of an optical pulse propagates in a medium. In general, the group velocity is the velocity of the peak of the pulse envelope (see the black dot in the dotted box of Fig. 12.4(a)). Readers might want to refer to Chapter 1 where the group velocity is initially defined.

Phase velocity: The velocity with which phase fronts propagate in a medium. In general, the phase velocity is the velocity of the peak of the electric-field carrier (see the filled triangle in the dotted box of Fig. 12.4(a)). Readers might want to refer to Chapter 1 where phase velocity is initially defined.

CEP: The phase between the peak of the optical carrier (field) and that of the envelope (see $\Delta\phi_{CE}$, in the dotted box of Fig. 12.4(a)). In a train of multiple pulses, CEP usually varies due to the difference between phase and group velocity. For a mode-locked laser without active stabilization of the pulse train, $\Delta\phi_{CE}$ is a dynamic quantity, which is sensitive to the perturbation in the laser.

CEP locking: Stabilization of the pulse train in a mode-locked laser, which will be introduced in Section 12.4. A mode-locked laser source with effective CEP locking can be taken as a *frequency comb*.

Frequency comb: Optical frequency comb is a light source with a spectrum containing not a continuum but a discrete pattern of sharp, narrow, equidistant laser lines in the frequency space (see Fig. 12.4(b) solid lines under the laser pulse). Frequency combs can be generated by the stabilization of the pulse train in a mode-locked laser. Optical

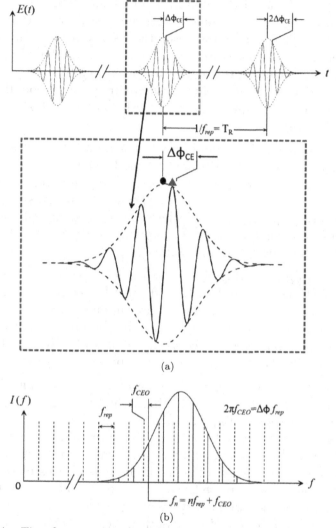

(a)

(b)

Fig. 12.4. Time-frequency correspondence and relation between the carrier-envelope phase (CEP, $\Delta\phi_{CE}$), repetition frequency (f_{rep}), and offset frequency (f_{CEO}). (a) In the time domain, the pulse train has a periodically pulse-to-pulse CEP shift $\Delta\phi_{CE}$ for the emitted pulses. The pulse envelop (dash) and the underlying electric-field carrier (solid) in the dotted box of the upper panel is enlarged in the down panel. (b) In the frequency domain, the longitudinal modes (solid) under the laser pulse (Gaussian in shape in this illustration) are offset from integer multiples (dash) of repetition frequency f_{rep} by an offset frequency f_{CEO}. Note: The horizontal axis in (a) is time. Although CEP shift $\Delta\phi_{CE}$ is labelled in (a) for convenience of explanation, please keep in mind that phase has no dimension.

frequency comb via CEP-stabilization of a femtosecond mode-locked laser is a Nobel-winning technique developed by Theodor Hänsch and John Hall in 1999. Both of them won the 2005 Nobel Prize in Physics "for their contributions to the development of laser-based precision spectroscopy, including the optical frequency comb technique".

We shall now briefly define absolute frequency, repetition frequency and off-set frequency here while the physics will be explained in the following text.

Absolute frequency: The absolute frequencies are the output series of frequencies from a CEP-stabilized mode-locked laser (see f_n in Fig. 12.4(b)). In a frequency comb, the absolute frequencies are equidistant frequency lines in optical spectrum. Usually, such kind of the absolute frequencies are associated with a regular train of ultra-short pulses, having a fixed pulse repetition rate which determines the inverse line spacing in the spectrum.

Repetition frequency: Repetition frequency is the frequency spacing of comb teeth (see f_{rep} in Fig. 12.4(b)), which is equal to the number of pulses emitted by a mode-locked laser in one second.

The pulse repetition rate (or pulse repetition frequency) f_{rep} of a regular train of pulses is defined as the number of emitted pulses per second, or more precisely the inverse temporal pulse spacing.

Offset frequency: The frequency separation between an absolute comb frequency and its nearest integer multiples of the repetition frequency (see f_{CEO} in Fig. 12.4(b)).

In the case of few-cycle pulses, the relative phase between the peak of the pulse envelope and the underlying electric-field carrier wave becomes relevant (see Fig. 12.4). In general, this phase is not constant from pulse to pulse because the group and phase velocities fluctuate inside the laser cavity (see Fig. 12.4(a)). In general, the electric field of a pulse including both the carrier and envelope can be expressed as:

$$E(t) = A(t)\cos(\omega_c t + \phi_{\mathrm{CE}}), \qquad (12.15)$$

where $A(t)$ is the envelope of the pulse, ω_c is the carrier frequency, and ϕ_{CE} is the absolute phase. The absolute phase is given by

$$\phi_{\mathrm{CE}} = \phi_0 + \Delta\phi_{\mathrm{CE}}, \qquad (12.16)$$

where ϕ_0 is an unknown overall constant phase, and $\Delta\phi_{\mathrm{CE}}$ is the phase between the peak of the optical carrier (field) and that of the envelope (carrier-envelope phase or CEP) (see Fig. 12.4(a)). Normally, due to dispersion, there is a difference between the phase velocity of the carrier and the group velocity of the envelope.

The carrier-envelop phase can be derived by accumulating the phase difference along the optical path,

$$\Delta\phi_{CE} = -\omega_c \int_0^{L_c} \left(\frac{1}{v_g} - \frac{1}{v_p}\right) dx = \left(\frac{1}{v_g} - \frac{1}{v_p}\right) L_c \omega_c, \quad (12.17)$$

where v_g and vp are the group and phase velocity, respectively, and L_c is the cavity length.

In a mode-locked laser, due to the difference between the phase velocity of the carrier (field) and the group velocity of the envelope, the carrier slips through the envelope when the pulse circulates in the cavity. Because an output pulse is only produced once per cavity round trip, the emitted pulse train from a mode-locked laser has an equal interval $T_R = 1/f_{\text{rep}}$ between two adjacent pulse envelopes in the time domain (see Fig. 12.4(a)). Without active stabilization of the pulse train, the value of CEP ($\Delta\phi_{CE}$) of a mode-locked laser is a dynamic quantity, which is different from pulse to pulse. Meanwhile, an optical Frequency comb can be from the output spectrum of a mode-locked laser through effective CEP stabilization (see Fig. 12.4(b)).

In order to look into the physics behind absolute frequency (f_n), repetition frequency (f_{rep}), and offset frequency (f_{CEO}), the electric field of the pulse train from a mode-locked laser is employed,

$$E_{\text{train}}(t) = \sum_m A(t-nT_R)\exp i(\omega_c t - n\omega_c T_R + n\Delta\phi_{CE} + \phi_0), \quad (12.18)$$

where m is an integer. In the frequency domain, the electric field of the pulse train is obtained through Fourier transform,

$$\tilde{E}_{\text{train}}(\omega) = \int \sum_n A(t - nT_R)\exp i(\omega_c t + n(\Delta\phi_{CE} - \omega_c T_R) + \phi_0)$$

$$\times \exp(-i\,\omega t)dt$$

$$= \sum_n \exp[i(n(\Delta\phi_{CE} - T_R) + \phi_0)] \int A(t - nT_R)$$

$$\times \exp[-i(\omega - \omega_c)t]dt$$

$$(12.19)$$

Considering $\tilde{A}(\omega) = \int A(t)\exp[-i\omega t]\mathrm{d}t$ and $\int f(X - X_0)\exp(-iax)\mathrm{d}x$
$= \exp(-iax_0)\int f(x)\exp(-iax)\mathrm{d}x$, Eq. 12.19 can be written as

$$
\begin{aligned}
\tilde{E}_{\text{train}}(\omega) &= \sum_n \exp[i(n(\Delta\phi_{CE} - \omega_c T_R) + \phi_0)] \\
&\quad \times \exp[-in(\omega - \omega_c)T]\tilde{A}(\omega - \omega_c) \\
&= \exp(i\phi_0)\bar{A}(\omega - \omega_c)\sum_n \exp[i(n(\Delta\phi_{CE} - \omega T_R))] \qquad (12.20) \\
&= \exp(i\phi_0)\bar{A}(\omega - \omega_c)\sum_m \delta(\Delta\phi_{CE} - \omega T_R - 2\phi m).
\end{aligned}
$$

where the last step is obtained by using the Poisson Summation Formula,

$$
\sum_m f(x - mp) = \sum_k \frac{1}{p}F\left(\frac{k}{p}\right)e^{2\pi ik/p}, \qquad (12.21)
$$

where $F(y)$ is the Fourier transform of $f(x)$. Therefore, with CEP-locking, the angular frequencies from a mode-locked laser are included in the last bracket of eq. (12.20). In this sense, we define

$$
\omega_n = \frac{2\pi n}{T_R} - \frac{\Delta\phi_{CE}}{T_R}. \qquad (12.22)
$$

In fact, in eq. (12.20) angular frequency (ω_n) can be derived from the absolute frequency by using $\omega_n = 2\pi f_n$, and $1/T_R$ equals to f_{rep}. Therefore, the relationship between absolute frequency f_n and repetition frequency (f_{rep}) is

$$
f_n = nf_{rep} - \frac{\Delta\phi_{CE}}{2\pi}f_{rep}, \qquad (12.23)
$$

The absolute frequency (f_n) is indicated in Fig. 12.4(b) by a series of solid lines. For convenience of explanation, $f_{\text{rep}}, 2f_{\text{rep}}, 3f_{\text{rep}}, \ldots, nf_{\text{rep}}$ (see eq. (12.23)), integer multiple of f_{rep}) are indicated by the series of dashed lines in Fig. 12.4(b). Besides, it can be found that the absolute comb frequencies f_n are not necessarily integer multiples of the repetition rate (see the second term on the right hand side of eq. (12.23)). The second term is defined as offset frequency f_{CEO}. Therefore, the absolute optical frequency is

$$f_n = n f_{rep} + f_{CEO}, \qquad (12.24)$$

where the absolute frequency f_n is also the n-th comb tooth. By comparing the second term at the right hand side of eq. (12.23) with eq. (12.24), it writes,

$$\Delta\phi_{CE} = -2\pi f_{CEO}/f_{rep}. \qquad (12.25)$$

In eq. 12.25, we can find that this offset is due to the phase difference between the peak of the optical carrier (field) and that of the envelope. Locking the CEP is also equivalent to locking the absolute optical frequencies of the comb, and vice versa.

Equation (12.25) is the key to answer the question at the beginning of this section "**How could we lock the CEP of a mode-locked laser?**". At the first glimpse, the CEP is due to the group and phase velocities difference from pulse to pulse inside the cavity, which is too abstract to stabilize. However, eq. (12.25) tells us that **CEP can be locked by locking both repetition frequency (f_{rep}) and offset frequency (f_{CEO})**. How to lock repetition frequency and offset frequency will be described in the next section.

12.4 Carrier-Envelope Phase Measurement and Stabilization

It is known from the previous section that the CEP can be locked by locking both the repetition frequency and offset frequency. The purpose of the current section is to introduce how we can measure and stabilize the CEP.

12.4.1 Carrier-envelope phase measurement

Equation (12.20) indicates that the CEP $\Delta\phi_{CE}$ can be derived from the repetition rate f_{rep} and offset frequency f_{CEO}. The repetition rate f_{rep} can be detected by employing the combination of a photoelectric detector and a spectrum analyzer (or a frequency counter). The offset frequency f_{CEO} can be measured by using f-to-2f self-referencing detection.

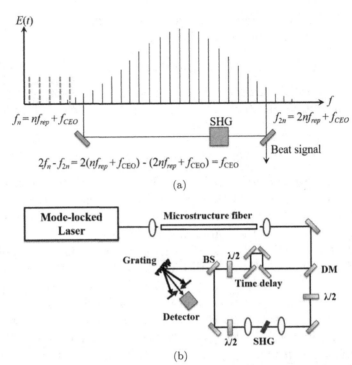

Fig. 12.5. f-to-$2f$ self-referencing detection in the frequency domain. The frequency comb is indicated by the 'rainbow' teeth. (b) The schematic diagram for f-to-$2f$ self-referencing detection. The frequency comb is generated by a mode-locked laser. BS: beam splitter; SHG: second harmonic generation; DM: dichroic mirror; APD: avalanche photodiode; $\lambda/2$: half-wave plate.

To explain the f-to-$2f$ self-referencing technique, we assume that the pulse to measure is an optical frequency comb. We also assume that the pulse is firstly spectrally broadened inside a nonlinear crystal or microstructure fiber. In this sense, octave-spanning spectra (continuous spectra which cover a region from the fundamental wavelength to its second harmonic) is generated (see Fig. 12.5).

In this case, the second harmonic of the low-frequency (red) end of the comb will overlap with the high-frequency (blue) end (see Fig. 12.5(a)). For the low-frequency (red) end of the comb, the phase for the electric field of the n-th comb tooth is given by:

$$\phi_n = 2\pi f_n t + \phi_{0n} = 2\pi(n f_{rep} + f_{CEO})t + \phi_{0n}, \qquad (12.26)$$

where ϕ_{0n} is a constant for the initial phase of the n-th comb tooth. Similarly, for the high-frequency (blue) end of the comb, the phase for the electric field of the $2n$-th comb tooth can be written as:

$$\phi_{2n} = 2\pi f_{2n}t + \phi_{02n} = 2\pi(2nf_{rep} + f_{CEO})t + \phi_{02n}.$$

(12.27)

The second harmonic of the low-frequency (red) end of the comb overlaps with the high-frequency (blue) end. Thus, the heterodyne beat between the second harmonic and fundamental combs produces a radio-wave frequency:

$$\phi_{detect} = 2\pi f_{CEO}t + 2\phi_{0n} - \phi_{02n}.$$ (12.28)

Therefore, the heterodyne beat can be detected by a photoelectric detector, and f_{CEO} can be derived from the waveform of the beat from an oscilloscope. Figure 12.5(b) is the experimental setup for measuring f_{CEO}. The octave-spanning spectrum is produced in the microstructure fiber. Then, the components of the low-frequency and high-frequency ends are separated by a dichroic mirror. The low-frequency part gets frequency doubled by a BBO crystal. After that, the two arms are recombined. The heterodyne beat from two optical frequency combs can be detected by an avalanche photodiode (APD). If the pulse spectral width is less than one octave, the $2f$-to-$3f$ self-referencing scheme can be applied to detect f_{CEO} as well.

Both f-to-$2f$ and $2f$-to-$3f$ self-referencing schemes can be used for f_{CEO} detection.

f-to-$2f$ self-referencing scheme:

$$2f_n - f_{2n} = 2(nf_{rep} + f_{CEO}) - (2nf_{rep} + f_{CEO}) = f_{CEO}.$$ (12.29)

$2f$-to-$3f$ self-referencing scheme:

$$3f_n - 2f_{3n/2} = 3nf_{rep} - 2 \times (3n/2)f_{rep} + 3f_{CEO} - 2f_{CEO} = f_{CEO}$$

(12.30)

For the same reason, the $3f$-to-$4f$ self-referencing scheme can be applied. Therefore, CEP ($\Delta\phi_{CE}$) can be derived from f_{rep} and f_{CEO} by applying Eq. (12.20).

12.4.2 Carrier-envelope phase stabilization

The CEP is a very sensitive parameter, reacting strongly to any fluctuations in a laser resonator. From the first part of this section, we know that CEP locking is to lock repetition frequency f_{rep} and offset frequency f_{CEO}. The drift of f_{rep} is mainly due to the fluctuations from the cavity length and refractive index. The cavity length is mainly affected by the mechanical vibration, while the refractive index of the gain medium is related with airflow, environmental temperature, humidity, pressure, etc. The drift of f_{CEO} is mainly due to the fluctuation of intracavity dispersion, the cavity length, and the refractive index. Besides, the fluctuation of the intracavity intensity of laser pulse can also lead to the drift of f_{CEO}.

Therefore, to lock CEP, the laser cavity is suggested to be built inside a super-clean laboratory with little air flow and constant temperature, humidity, pressure, etc. Besides, to minimize temperature fluctuation and mechanical vibration, the laser cavity can be made from materials with a low thermal expansion coefficient and large shock-wave absorption rate. Meanwhile, to minimize the influences from the airflow and the pressure, the laser cavity should be covered and packaged.

CEP locking in a laser oscillator can be done by actively locking the cavity length and intracavity dispersion.

A simplified schematic diagram for active CEP locking in a Ti:sapphire oscillator is shown in Fig. 12.6. The "Carrier-envelope phase locking electronics" in Fig. 12.6 actually consists of frequency

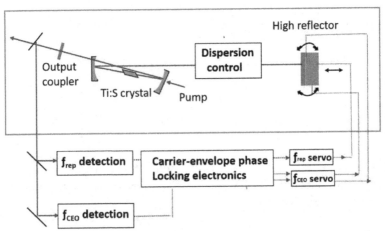

Fig. 12.6.   Experimental setup for CEP locking. Solid lines are optical parts and dashed lines are electrical paths. The high reflector mirror is mounted on a transducer to provide both tilt and translation.

tracking and feedback loops. In Fig. 12.6, a Ti:sapphire oscillator can be mode-locked by compensating the normal dispersion by the instruments for dispersion control (e.g. a pair of prisms). The high-reflector mirror is mounted behind the prism pair on a piezoelectric transducer tube (PZT) that allows both tilt and translation.

Here, f_{rep} can be measured by a spectrum analyzer or a counter. f_{CEO} detection is done by the f-to-$2f$ self-referencing scheme, which is introduced in the first part of this section. By comparing the detected f_{rep} with a constant radio frequency from a radio frequency synthesizer, a feedback loop can lock f_{rep} at the radio frequency. By translating the high-reflector mirror, the optical length of the cavity is locked to a constant value (see Fig. 12.6). Since in a solid laser oscillator, mode locking can be done by adjusting the group velocity dispersion by a pair of prisms, tilting the mirror after the prism pair provides a linear phase change with frequency. In this sense, f_{CEO} can be locked through programming the PZT by tilting the high-reflector mirror in order to eliminate the drift of f_{CEO} (see Fig. 12.6). Therefore, according to Eq. (12.20), the CEP can be stabilized with locked repetition frequency and offset frequency.

12.5 Closing Remark

In this chapter, mode locking and carrier-envelope phase (CEP) locking are introduced. Mode locking helps to decrease the output pulse duration of a laser to a picosecond or femtosecond region. Optical frequency combs via CEP-stabilization of femtosecond mode-locked laser is a Nobel winning technique, due to the simplicity of the technique to obtain very precise spectroscopic data for various calibration as compared to traditional technique. For instance, optical frequency metrology by CEP-locking can determine the unknown frequencies with extreme precision. It is preferred by fundamental physical constants calibration (e.g. fine structure constant and charge-mass ratio), molecular/atomic transition measurements (e.g. Rydberg constant and lamb shift), quantum electrodynamics investigation, etc. Besides, CEP-locked lasers have also demonstrated their applications on gravity wave detection, optical communication, optical ranging, etc.

Chapter 13

Chirped Pulse Amplification

In recent years, tremendous developments in the physical world have been made due to the advent of high-power lasers, including modern multi-petawatt (PW, 10^{15}W) lasers. One of the most important technology behind this is chirped pulse amplification (CPA). CPA is the 2018 Nobel physics prize-winning technology invented by Gérard Mourou and Donna Strickland. It has rapidly empowered both research institutions and universities to set up table-top high power laser systems from its initial modest demonstration to multi-terawatt (TW: 10^{12} W) and multi-peta-watt scale systems. Numerous scientific, industrial and medical applications have been advanced based on this new laser technology. This chapter includes six sections: *The Nobel Prize in Physics 2018, Race for the petawatt, CPA concept, Pulse stretching and compression, Regenerative amplification, and Multi-pass amplification.*

13.1   The Nobel Prize in Physics 2018

The Nobel Prize, first awarded in 1901, is widely considered as the highest honor in science, economics, literature, etc. The Nobel Prize in Physics in 2018 was partially granted to Gérard Mourou and Donna Strickland "for their method of generating high-intensity, ultra-short optical pulses". The method is called CPA technology.

As we introduced in previous chapters, with the advent of the laser, the output pulse duration has continuously decreased from mircosecond with free running laser, to nanosecond with Q switching, and finally to picosecond and femtosecond with mode-locking. Mode locking has shortened the pulse duration to a few femtoseconds. A shorter pulse duration usually corresponds with higher peak intensity and larger frequency coverage.

Historically, the amplification of a femtosecond pulse came along with two technical "challenges". The first is that the amplifying medium must have a large bandwidth, which accommodates the frequency coverage of the seed pulse. The developments of modern laser material sciences can help to solve this problem by producing pretty good gain media for laser amplifiers, for instance, crystals of Nd:YAG, Ti:sapphire, CaF_2, etc., which have large gain bandwidth (tens of nanometers or even broader), high-energy storage, and a short lifetime of the lower excited state for the transition. The second is that during high power amplification, the seed pulse usually works under a saturation fluence condition. Based on this condition, the intensities in amplifiers are above the limit for preventing nonlinear effects and optical damage inside the amplifying medium and optical components. Especially, when the laser peak power in the amplifiers exceeds several GW (giga-watt or 10^9 watts), the laser beam will encounter spatial intensity modulations. The resulting self-focusing effect can break up the beam into multi-narrow high intensity traces called filaments. Such filaments could damage the amplifying medium or other optical components. This means that such a drawback of unavoidable damages prevented the peak power to increase further beyond the GW level.

The revolution happened in 1985. Laser physicists at the University of Rochester demonstrated the CPA technique, which overcame the above mentioned amplification issues. To avoid unhealthy intensities in the optical amplifier, the CPA technique stretched the pulse duration of the seed pulse in the time domain by the chirping technique (see below for a detailed explanation). Then, the peak power of the pulse was reduced with unchanged laser fluence. The chirped pulse was amplified afterward, and finally, the amplified pulse was compressed back into a short but now energetic pulse. Therefore, the output pulse exhibited both strong pulse energy and short duration.

In the first CPA setup, the laser pulse was stretched using an optical fiber with a positive dispersion and was recompressed by a pair of gratings, with a negative group delay dispersion.

Although non-negligible high order dispersion usually resulted in the propagation of a femtosecond pulse in an optical fiber, this first embodiment led to a spectacular 100-fold improvement in peak power.

13.2 Race for the Petawatt

CPA is now a popular laser technology and has paved the way for higher-intensity, ultra-short laser pulse generation. According to the *Nobel Lecture: Extreme light physics and application* by Gérard Mourou, CPA revolutionized the physical world in three ways:

(1) CPA technique could help build TW (terawatt, 10^{12}W) systems that fit on a tabletop, delivering intensities 10^5–10^6 times higher than utilizing conventional technology.
(2) CPA architecture could be easily retrofitted to existing large laser fusion systems at a relatively low cost.
(3) CPA lasers could be combined with large particle accelerators due to their compact size.

Besides, the progress of the CPA technique has inspired the OPCPA technique (OPCPA: Optical Parametric Chirped Pulse Amplification), capable of generating sub-100 fs, mid-infrared energetic pulses. For instance, the OPCPA system in the Vienna University of Technology can deliver sub-100 fs laser pulses with a central wavelength of 3.9 μm and maximum pulse energy of 30 mJ at a 20 Hz repetition rate. This OPCPA system has been used to generate high-order harmonics and terahertz pulses with very high energy.

On one hand, due to the generated ultrashort pulse duration, CPA is strongly connected with our daily life by producing clean cuts of minimal roughness even at the atomic scale. This has provoked the applications of photolithography, which can integrate microelectronics into a tiny piece of wafer. In the past 20 years (2000–2020), the best resolution of modern lithography machines has decreased from hundreds of nanometers to several nanometers using CPA lasers. Besides, in medical sciences, CPA lasers have given rise to applications in ophthalmic procedures such as refractive surgery, cataract surgery, corneal transplant, and glaucoma. Today, millions of patients a year are benefiting from CPA femtosecond laser interventions. On the other hand, the generated ultra-strong intensity stimulates people to look forward in particle production in "empty" spaces as well as laboratory astrophysics and cosmology, nuclear optics, neutron physics, and attosecond (10^{-18} second) physics, etc.

Nowadays, the CPA technique stirs up worldwide competition for the construction of super-powerful lasers. Starting from 2006, initially initiated by Gerard Mourou, the extreme light infrastructure (ELI) program in the European Union (EU) has been proposed by $\sim$40 research institutions from more than 10 EU countries in order to develop ultrahigh peak power (petawatts) laser devices. The ELI program has been expected to create a new era for the research and application of the interaction between laser and matter. Besides, the United Kingdom (UK), France, China, Russia, Japan, South Korea and India have also put forward similar research plans for the construction of ultrahigh peak power laser beamlines.

(1) ELI program

The ELI program aims to operate one of the world's most intense laser systems. It offers a unique source of radiation for particles acceleration, which enables pioneering research not only in physics but also in biomedicine and laboratory astrophysics and many other fields. There are four facilities under the ELI program, namely, ELI-Beamlines Facility (Prague, Czech Republic), ELI-Attosecond Facility (Szeged, Hungary), ELI-Nuclear Physics Facility (Magurele, Romania), and ELI-Ultra High Field Facility (not addressed yet as of August 2021). ELI has successively started the construction of the first three devices, with a total investment of 850 million euros. It also has successfully constructed a series of high-power laser systems, including 2×10 PW @ 1 shot per minute, 2×1 PW @ 1 Hz, 2×100 TW @ 10 HZ pulses, 0.5 PW @ 3.3 Hz (pulse duration of < 30 fs), etc. The 10-petawatts lasers have been used for γ-ray generation.

(2) XCELS program

Exawatt Center for Extreme Light Studies (XCELS) program is located in Russia. It was proposed by the Institute of Applied Physics of the Russian Academy of Sciences. The XCELS program aims to realize 200-petawatts lasers, which will be constructed by coherently combining 12 laser beams; each beam produces 15 PW/25 fs pulses. This facility is planned to carry out exploration on the space-time structure of high-energy physics, strong field physics in a vacuum, as well as laboratory astrophysics and cosmology, nuclear optics, and attosecond physics, etc.

(3) Programs in the United States

The Berkeley Lab laser accelerator (BELLA) program is carried out by the Lawrence Berkeley National Laboratory. In the BELLA program, a petawatt laser with 40-J pulse energy, 40-fs pulse duration, and 1-Hz repetition rate has been constructed. In the future, 1 TeV level electron-position collider with plasma accelerators can be expected. In addition, the University of Rochester, Lawrence Livermore National Laboratory, the University of Texas, and the University of Nebraska Lincoln have also exhibited their blueprints for developing their petawatt laser systems.

(4) The CLF in UK

The Central Laser Facility (CLF) of Rutherford Appleton Laboratory in the UK offers researchers in the UK and other EU countries high-power neodymium glass laser devices, Ti:sapphire laser devices, and several small-scale lasers to carry out research in material sciences and atomic physics. Among them, the biggest laser device is called the Vulcan laser. Vulcan laser is a petawatt laser system based on CPA technique. It is composed of Nd:glass amplifier chains capable of delivering up to 1 PW peak power in 500-fs duration at 1053 nm.

(5) Apollon system in France

The Apollon system, which was built by the French National Center for Scientific Research, Paris Polytechnic University, and French Institute of Higher Science and Technology, is located in Paris. The CPA-based system can realize 300 J for the direct output pulse energy before its compressor. After compression, 150-J pulse energy, 15-fs pulse duration, and 10-petawatt peak power can be expected according to the design. Its main applications include ion and electron acceleration, X-ray generation, and strong field physics.

(6) Programs in China

The Shanghai Superintense Ultrafast Laser Facility (SULF) is located in Shanghai, China. It was built by the Shanghai Institute of Optics and Fine Mechanics (SIOM). The SULF-10 PW laser beamline is based on the double CPA scheme. The amplifier output energy is $\sim$408 J before the compressor. Compressed pulse duration of the amplified laser and the total transport efficiency for compression are 22.4 fs and 70.52% respectively. In addition, the Laser Fusion Center and Institute of Physics of Chinese Academy of Engineering Physics, Shanghai Jiaotong University, Peking University, and China Institute of Atomic Energy have built or projected to construct their femtosecond laser system with a peak power of beyond 1 petawatt.

(7) Program in South Korea

In the Center for Relativistic Laser Science (CoReLS), Institute for Basic Science, Gwangju, South Korea, two PW beamlines have been developed, one at a 1 PW/20 fs @ 0.1 Hz repetition rate and the other at a 4 PW/20 fs @ 0.1 Hz repetition rate. This laboratory is rather unique in the sense that many interesting interaction experiments are consistently being carried out in parallel to the development of the PW laser beam lines over the past almost ten years. In 2021, a record breaking ultra-high intensity of 1.1×10^{23} W/cm^2 was obtained in this laboratory through wave front correction and tight focusing. This achievement is a world record so far. Around and beyond this intensity, a new regime of relativistic laser-matter interaction physics is being explored in this laboratory such as the study of strong field quantum electrodynamics by enabling exploration of nonlinear Compton scattering and BreitWheeler pair production. Relativistic high harmonics generation is also being explored.

(8) India
In India, a PW laser facility has been planned for construction aiming to be operational in about two years time or less.

13.3 CPA Concept

In order to avoid detrimental nonlinear pulse distortion or destruction of the gain medium or of other optical elements, the pulses from the laser oscillator in a CPA beamline are chirped and temporally stretched to a much longer duration in a stretcher before entering an amplifier according to group dispersion described in Section 1.5. The schematic diagram is illustrated in Fig. 13.1. The pulse duration is extended in a stretcher. The long pulse duration reduces the peak

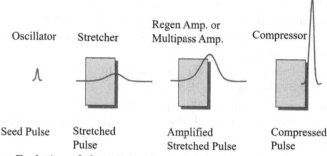

Fig. 13.1. Evolution of the temporal pulse shape in a chirped-pulse amplifier. Amp: amplifier.

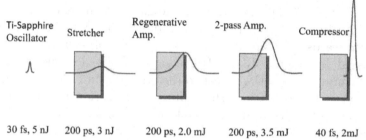

Fig. 13.2. Evolution of the temporal pulse shape in a Ti-sapphire femtosecond cascade laser system. Typical pulse durations and pulse energies are shown at the bottom line. Amp: amplifier.

power to a level where the above-mentioned detrimental effects in the gain medium of an amplifier are effectively prevented.

After the amplifier, a dispersive compressor is used to remove the chirp and temporally compress the pulses to a duration similar to the input pulse duration. In Fig. 13.1, the compressor narrows the pulse duration, which is similar to that of the oscillator, while the peak intensity is much higher.

To show a concrete example, we shall now analyze in detail a rather popular commercial CPA-based Ti-sapphire femtosecond cascade laser amplification system with central wavelength around 800 nm. Figure 13.2 shows a schematic diagram of a typical CPA-based Ti-sapphire femtosecond cascade laser amplification system with central wavelength around 800 nm. The cascade amplification system consists of one stretcher, one regenerative amplifier, one 2-pass amplifier, and one compressor. The input pulse for the amplification system is actually the output from a mode-locked oscillator, which produces pulses with 30-fs pulse duration, 5-nJ pulse energy, and 40-nm spectral width.

The input beam from the oscillator first enters the stretcher (detail to be discussed in Section 13.4). The pulse duration is stretched from 30 fs to ∼200 ps in the stretcher, while the pulse energy is reduced to ∼3 nJ due to ∼40% diffraction loss in the stretcher (see Fig. 13.2).

The stretched pulse then goes through the regenerative amplifier (to be discussed in Section 13.5) and the 2-pass amplifier (to be discussed in Section 13.6). The pulse oscillates and gets amplified in the regenerative amplifier. The 2-pass amplifier is placed after the regenerative amplifier for the sake of achieving even higher peak power. The maximum pulse energy is now around ∼3.5 mJ (see Fig. 13.2).

Afterwards, the beam is sent to the compressor (detail to be discussed in Section 13.4). The compressor is designed to be at phase conjugation with respect to the stretcher in order to minimize the high order dispersion. Here, phase conjugation means that the stretcher and the compressor match over all orders of dispersion but opposite in sign. This will be explained in Section 13.4. The output pulses from the CPA-based device have 40-fs pulse duration and up to 2 mJ pulse energy. The details for pulse stretching and compression, regenerative amplification, and multi-pass amplification will now be discussed in the following sections.

13.4 Pulse Stretching and Compression

Several stretcher and compressor implementations have been applied in the CPA system in order to stretch or compress the pulse duration by chirping techniques. Appropriate stretching and compression ratios can be achieved with grating pairs, prism pairs, chirped mirrors, optical fibers, fiber Bragg gratings, etc. We shall use a grating pair to illustrate the principle of stretching and compression. Generally speaking, when injecting a pulse into a stretcher the shorter wavelength components of the pulse see longer optical paths than the longer wavelength components (to be explained later). The result is that the longer wavelength components exit the stretcher first.

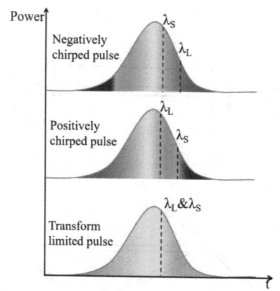

Fig. 13.3. Schematic diagram showing a negatively chirped pulse, a positively chirped pulse, and a transform limited pulse.

Hence, the pulse duration is expanded or stretched. A stretcher can stretch pulses from femtosecond pulse duration to picosecond and nanosecond durations. With a similar principle to the stretcher, the compressor is normally designed to make the shorter wavelength components of the above mentioned stretched pulse exit first. This would compress the pulse duration.

Figure 13.3 shows schematically a positively chirped pulse with shorter wavelength lagging behind the longer wavelength and a negatively chirped pulse with shorter wavelength ahead of longer wavelength. A transform limited pulse having all the wavelengths superposed on top of one another temporally would be the result after compression.

In general, both the stretcher and the compressor provide a certain number of multiple orders of dispersion. In a practical system, the number of multiple orders of dispersion offered by the stretcher and the compressor are equal, but opposite in sign. The condition is that the grating pairs in the stretcher and the compressor have the same grating groove density, orientation, and separation.

In this section, the Treacy-Martinez grating arrangement is introduced. The Martinez grating pair is a stretcher (see Fig. 13.4(a)),

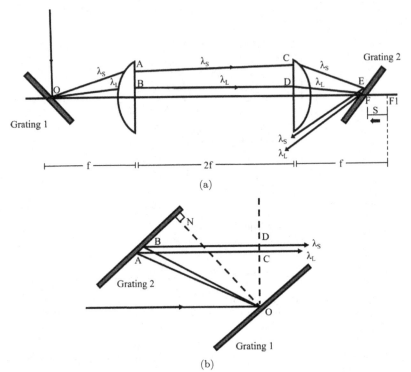

Fig. 13.4. (a) A pulse stretcher using the Martinez grating arrangement. The two identical gratings are anti-parallel to each other. (b) A pulse compressor using the Treacy grating arrangement. The two identical gratings are parallel to each other. They are identical to the grating pair in (a). f is the focal length of the identical lenses. S is the displacement of the second grating towards the left hand side from the geometrical focus (F1) of the second lens (see the arrow). λ_L and λ_S indicate the long and short wavelength components, respectively. In (b), we define the perpendicular distance between the gratings ON as G, the slant distance OA (or OB) as b.

while the Treacy grating pair, a compressor (see Fig. 13.4(b)). Nowadays, (in 2021), such an arrangement is used in most commercial CPA systems as well as in many terawatt and petawatt laser setups.

13.4.1   Martinez grating pair (laser pulse stretcher)

The Martinez grating pair offers positive group chirp. It is often used to stretch the pulse from the oscillator. The original arrangement of a Martinez grating pair is shown in Fig. 13.4(a). It was proposed by Martinez from the Rochester group. A Martinez grating pair is generally designed to be in phase conjugation with a Treacy grating pair (see later); i.e. both of them have the same grating groove density, orientation, and separation. As shown in Fig. 13.4(a), the input pulse from a mode-locked oscillator is normally a transform limited laser pulse; i.e. all the wavelength/frequency components in the pulse propagate together; i.e. in-phase (see Fig. 13.3). This pulse is first incident on Grating 1 at point O (see Fig. 13.4(a). Due to different diffraction angles, the short wavelength component (λ_S) and the long wavelength component (λ_L) are spatially separated; they go through the grating pair through the light paths OACE and OBDF, respectively. Here, points A and B are at the first lens, points C and D at the second lens and points E and F at the surface of the second grating (Grating 2). Grating 2 is at the left hand side of the geometrical focus of the second lens, thus, positive chirp (see Fig. 13.3, shorter wavelength lagging behind longer wavelength components in the output pulse) can be induced by this arrangement (path OACE is longer than path OBDF). Thus, at the output of the second grating, because path OACE is longer than path OBDF, longer wavelength components are ahead of the shorter wavelength components in the output pulse from the second grating. We define this pulse as being positively chirped (see Fig. 13.3). Therefore, the duration of a pulse (from the oscillator) can be stretched in a Martinez grating pair (laser stretcher).

As shown in Fig. 13.4(a), the arrangement of a Martinez grating pair is based on a $4f$ optical system including two gratings and two lenses. In a $4f$ optical system, two identical lenses are used. The separation distance between the two lenses is $2f$ (f being the focal length of the lenses).

A $4f$ optical system is based on the Fourier transform theory. In a $4f$ optical system, a plane wave is assumed incident from the left. A electrical field distribution in the beam cross section containing one 2D function, f(x,y), is at the input plane of the system, located one focal length at the left hand side of the first lens in Fig. 13.4(a). Here, x and y are coordinates at the beam cross section. The first lens spatially modulates the incident plane wave in both magnitude and phase. It produces an "image" (Fourier transform of f(x,y)) one focal length on the right of the first lens. The "image" also lies in the input plane of the second lens (one focal length on the left hand side of the second lens), so that the origin electrical field distribution f(x,y) at the input plane of the system is retrieved in the right focal plane of the second lens (f(x,y) Fourier transform twice). If an ideal, mathematical point source of light is placed at the focal point in the input plane of the first lens, then there will be a uniform, collimated field produced in the output plane of the first lens. This uniform, collimated field is retrieved by the second lens at the right focal plane of the second lens (imaging plane of the system).

In a Martinez grating pair arrangement, the first grating (Grating 1) is placed at the input plane of the system (see Fig. 13.4(a)). The first lens collimates the diffracted light from Grating 1. The second grating (Grating 2) is placed at the left hand side of the imaging plane of the system (see Fig. 13.4(a)). In this sense, the arrangement can provide positive chirp. The advantage of the $4f$ optical system is that Grating 2 can retrieve both magnitude and phase information of the electrical field at Grating 1.

13.4.2 Treacy grating pair (laser pulse compressor)

The Treacy grating pair offers a negative chirp (see Fig. 13.3, shorter wavelength ahead of longer wavelength components). It is often used to compensate for the positive chirp from the stretcher so as to compress the pulse. In a Treacy grating pair arrangement, two gratings are placed in parallel and separated (see Fig. 13.4(b)).

The input positively chirped laser pulse (longer wavelength component ahead of shorter wavelength component) is sent onto the first grating (Grating 1) at the position O (see Fig. 13.4(b)). The diffracted short wavelength component (λ_S) and long wavelength component (λ_L) are spatially separated because of the different diffraction angles of Grating 1. The short wavelength component (λ_S) and the long wavelength component (λ_L) arrive at the second grating (Grating 2) at points A and B. After that, the output beam from Grating 2 is still parallel with spectral components spatially separated. The dash line OCD in Fig. 13.4(b) is perpendicular to the initial light path (before frequency separation).

In Fig. 13.4(b), the short wavelength component (λ_S) and long wavelength component (λ_L) go through the grating pair through the paths OBD and OAC, respectively. OBD is longer than OAC. Thus, the short wavelength component (λ_S) is ahead of the long wavelength component (λ_L) after passing through the grating pair. In this sense, negative chirp is imposed onto the initially positively chirped pulse. Therefore, the duration of an initially positively chirped pulse can be compressed in a Treacy grating pair (laser compressor), ideally back to a transform limited pulse (see Fig. 13.3).

13.5 Regenerative Amplification

There are basically two methods for ultrashort pulses amplification: regenerative amplification and multi-pass amplification. Laser amplifiers built on the technique of regenerative amplification are called regenerative amplifiers. Regenerative amplifiers utilize Pockels cells to adjust the injection and output time of oscillating pulses. Regenerative amplifiers generally have two different schemes: the *quarter-wave-voltage scheme* (*quarter-wave voltage regenerative amplifier*) and the *half-wave-voltage scheme* (*half-wave voltage regenerative amplifier*), which can be distinguished by the voltage applied to the Pockels cell. The structures of these two regenerative amplifiers are described in this section.

Quarter-wave-voltage regenerative amplifier

A quarter-wave-voltage regenerative amplifier is usually applied when the output pulse energy is "not so strong" (e.g., under 2 mJ).

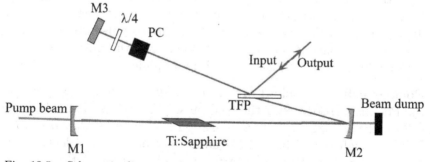

Fig. 13.5. Schematic diagram of a regenerative amplifier with a quarter-wave voltage applied to the Pockels cell. The input pulse is a stretched pulse usually of the order of 100 to 200 ps.

Note: PC: Pockels cell, TFP: thin film polarizer. $\lambda/4$: quarter-wave plate. M1 and M2 are concave mirrors. M3 is a flat reflective mirror.

The cavity of a quarter-wave voltage regenerative amplifier is a resonator. It means that the injected seed pulses can oscillate inside the cavity for several loops (dozens of round trips) until they gain enough power. Figure 13.5 shows a typical scheme of a quarter-wave voltage regenerative amplifier with Ti:Sapphire crystal as the gain medium being pumped by another laser at the green wavelength. Here, the quarter-wave voltage means that once the Pockels cell is switched on (with voltage applied), it is equivalent to a quarter-wave plate. Once the Pockels cell is off (without voltage applied), it can be taken as an isotropic crystal i.e. the laser beam propagates along the optic axis of the crystal inside the Pockels cell. The TFP (thin film polarizer) reflects vertically (perpendicular to the surface of the paper) polarized lights while transmits horizontally (parallel to the surface of the paper and perpendicular to the input beam) polarized ones. Both the entrance and exit for the seed beam are through the TFP.

The operations for a quarter-wave-voltage regenerative amplifier can be distinguished by whether the Pockels cell is switched on.

Pockels cell is switched off: When the Pockels cell is switched off, the incident vertically polarized seed pulse enters the cavity from the TFP. It will become horizontally polarized by passing through (back and forth) the quarter-wave plate twice (see Fig. 13.5). Then, the horizontally polarized oscillating pulse is able to pass through the TFP to the other end of the cavity and backtrack due to reflection by M1. After that, by passing through the quarter-wave plate (back and forth), the pulse will change its polarization direction to a vertical direction again. Finally, the vertically polarized pulse will emit from the resonator once it sees the TFP. Therefore, the pulse passes the gain medium twice and is only slightly amplified.

Pockels cell is switched on: A quarter-wave voltage is applied to the Pockels cell soon after the seed pulse passes through the Pockels cell for the second time. Its polarization is now horizontal. It propagates through the TFP, M2, M1 and comes back to the Pockels cell which is now equivalent to a quarter wave plate. It passes through the combination of the Pockels cell (with a quarter-wave voltage) and the quarter-wave plate; it is equivalent to passing through a half-wave plate (one quarter-wave plus another). After reflection by M3, the pulse passes through the combination of the quarter-wave plate and

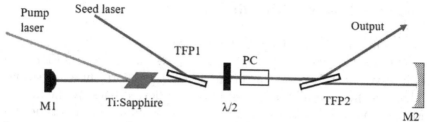

Fig. 13.6. Schematic diagram of a regenerative amplifier with a half-wave voltage applied to the Pockels cell. M1 is a convex mirror and M2 is a concave mirror. PC: Pockels cell, TFP: thin film polarizer.

the Pockels cell again. The round trip is equivalent to passing twice the combination. This can be taken as passing through a full wave plate. In this case, the pulse will not change its polarization (always horizontal polarization) during the oscillation inside the resonator. Therefore, the seed pulse is trapped and amplified in the resonator. Finally, after dozens of round trips, the intensity of the oscillating pulse in the resonator reaches a maximum. We can export the pulse by simply getting rid of the voltage on the Pockels cell. At this time, the polarization direction of the amplified pulse will change back to a vertical direction and therefore be reflected outside the cavity from TFP.

Half-wave-voltage regenerative amplifier

A half-wave-voltage regenerative amplifier is usually applied when the output pulse energy is "relatively strong" (e.g., 2–10 mJ). The cavity of a half-wave voltage regenerative amplifier is a resonator as well. Figure 13.6 shows the schematic diagram of a half-wave voltage regenerative amplifier. Ti:sapphire is used as the gain medium. The half-wave voltage means that once the Pockels cell is on (with voltage applied), it is equivalent to a half-wave plate. Once the Pockels cell is off (without voltage applied), it can be taken as an isotropic crystal; i.e. the laser beam propagates along the optic axis of the crystal inside the Pockels cell. There are two TFPs in Fig. 13.6. One (TFP1) is the entrance for the seed pulse, and the other (TFP2) is the exit for the amplified pulse.

The operation for a half-wave-voltage regenerative amplifier is described as follows.

Pockels cell is switched off:

When the incident vertically polarized seed pulse (usually of the order of 100 to 200 ps) is reflected into the resonator from TFP1, the half-wave plate rotates the polarization to a horizontal direction. Then, the horizontally polarized pulse is reflected back by the mirror M2 and passes through the half-wave plate again. At this time, its polarization direction changes back to a vertical direction. Therefore, the pulse will be exported from TFP1 without gain.

Pockels cell is switched on:

Same as the previous case, in the resonator, the incident vertically polarized seed pulse changes its polarization to horizontal direction by the half-wave plate and passes through the Pockels cell. Then, in this case, a half-wave voltage is immediately applied to the Pockels cell soon after the horizontally polarized pulse passes through the Pockels cell. After the pulse is reflected by M2, it sees a new combination of the Pockels cell and the half-wave plate being equivalent to a full wave plate. It means that the pulse will maintain its polarization along a vertical direction. Thus, the pulse is trapped inside the cavity and is amplified through multiple reflections. Finally, the energy of the seed pulse in the resonator reaches a maximum after dozens of round trips. We can export the amplified pulse by simply switching off the voltage on the Pockels cell. Therefore, the amplified pulse will rotate its polarization to a vertical direction and can be reflected outside the cavity from TFP2.

A regenerative amplifier is usually of high efficiency due to the resonant cavity. The amplification factor in a regenerative amplifier can reach 10^6–10^7. The pump-to-laser efficiency can be more than 20%. Besides, the intensity of the <u>transverse mode</u> from the output beam of a regenerative amplifier is a Gaussian mode (see Section 5.9 for the definition of the transverse mode). In a regenerative amplifier, it is necessary to maintain higher intensity (small diameter of the beam) in the gain medium, while maintaining lower intensity (relatively large diameter of the beam) at other optical instruments. In general, the gain medium in a regenerative amplifier is selected to own a high damaging threshold (5–$10\,\text{GW/cm}^2$).

13.6 Multi-pass Amplification

We learnt in the previous section that a regenerative amplifier is of good beam quality and has high pump-to-laser conversion efficiency. However, the laser beam has gone through many loops inside the cavity before energy saturation. In each loop, the beam passes through the gain medium, Pockels cells, and other polarization-control instruments. When the energy (hence peak power) of the pulse becomes higher and higher, it could bring in irreparable high order dispersion. The additional high order dispersion increases the difficulties for compression, thus, hard for few-cycle pulses generation. The reason is that the compressor and the stretcher match all linear orders of dispersion in the Treacy-Martinez grating arrangement. With additional dispersion introduced by the amplifier, the compressor and the stretcher are not perfectly matched any more. Fortunately, some mismatching of the second order dispersion (low order dispersion) can be optimized by adjusting the separation distance between the grating pairs of the compressor, because grating pairs provide mainly the second order dispersion. In this sense, high order dispersion is still left behind in the pulse.

The problem can be solved by multi-pass amplification with less intracavity instruments. A multi-pass amplifier is geometrically arranged for multiple passes of a beam through an amplifier (see Fig. 13.7, to be analyzed in the following paragraphs). The simplest case is that of a 2-pass amplifier, where a beam passes through the crystal twice with opposite propagation directions. There is no oscillation inside a multi-pass amplifier, and this means that the beam is "buying a one-way ticket" through the multi-pass amplifier. The amplified pulses see no Pockels cells or other dispersive elements except for the gain medium in a multi-pass scheme.

Figure 13.7 illustrates the schematic diagram of a 4-pass amplifier. The gain medium is a Ti:Sapphire rod, which is pumped in two counter-propagating directions. The seed beam is folded and reflected back by mirrors in order to pass through the Ti:sapphire rod four times. The energy of the seed beam is amplified by the inverted population at the Ti:sapphire rod each time it passes through the rod.

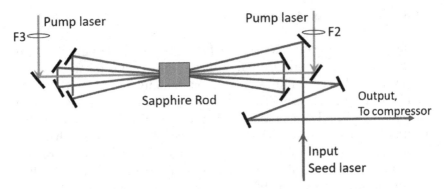

Fig. 13.7.   Schematic diagram of a 4-pass amplifier. F1, F2 and F3 are lenses. *Note*: PC: Pockels cell; $\lambda/2$: half-wave plate.

Spatial filters can be implemented for optimizing the beam quality. A multi-pass amplifier is able to deliver high-power picosecond-femtosecond pulses with pulse energy in the range of hundreds of Joules.

We introduced the regenerative amplifier and multi-pass amplifier in Sections 13.5 and 13.6, respectively. Whether to use multi-pass amplifiers or regenerative amplifiers depends on the requirements of the compressed pulse. In general, regenerative amplifiers can produce pulses which can be compressed to a duration of more than 20 fs at a relatively low energy (a few mJ). For the few-cycle pulse ($<20\,\text{fs}$) generation multi-pass amplifiers can be used. In the case of high-power laser systems based on CPA requiring very high pulse energy (at multi-joule level), these are usually mixed, that is, regenerative amplifiers for pre-amplifiers, while a multi-pass amplifier for the main amplifier.

13.7   Closing Remarks

In general, the CPA system includes a femtosecond mode-locked oscillator, a pulse stretcher, regenerative amplifier, multi-pass amplifier, and pulse compression. The pulse stretcher extends the pulse duration of the seed pulse in the time domain by the chirping technique. The chirped pulse is amplified afterward. The regenerative amplifier and multi-pass amplifier are two typical amplifiers. Finally, the amplified pulse is compressed back into a short but now energetic pulse at the pulse compressor. Therefore, the output pulse exhibited both strong pulse energy and short duration.

Nowadays, (in 2021), the tabletop Ti:sapphire femtosecond laser system has become a well-known commercial product. That is why we introduce Ti:sapphire-based amplifiers when we talk about laser amplifiers. Besides, the Yb(Er)-doped fiber laser system is another frequently used commercial CPA product, which will not be discussed here.

Index

345

Printed in the United States
by Baker & Taylor Publisher Services